Technologies for Development

Silvia Hostettler · Samira Najih Besson
Jean-Claude Bolay
Editors

Technologies for Development

From Innovation to Social Impact

UNESCO Chair in
technologies for development
Lausanne (Switzerland)

Editors
Silvia Hostettler
Ecole Polytechnique Fédérale de Lausanne
Lausanne
Switzerland

Samira Najih Besson
Cooperation & Development Center
Ecole Polytechnique Fédérale de Lausanne
Lausanne
Switzerland

Jean-Claude Bolay
Cooperation & Development Center
Ecole Polytechnique Fédérale de Lausanne
Lausanne
Switzerland

ISBN 978-3-030-08169-0 ISBN 978-3-319-91068-0 (eBook)
https://doi.org/10.1007/978-3-319-91068-0

Printed on acid-free paper

This Springer imprint is published by the registered company Springer International Publishing AG part of Springer Nature
The registered company address is: Gewerbestrasse 11, 6330 Cham, Switzerland

Foreword

The contribution of innovation and technology to sustainable development was at the heart of the 2016 edition of the Conference on Technologies for Development (Tech4Dev), organized by the UNESCO Chair in Technologies for Development at the Ecole Polytechnique Fédérale de Lausanne (EPFL). Beyond the importance of technological innovation for sustainable development, this Conference raised a question that appears crucial to UNESCO in order to respond adequately to today's complex economic, societal, environmental, and cultural challenges: how do we get from innovation to social impact?

In order to maximize the overall positive benefits of science, we need to *incorporate a vision of innovation in Science, Technology and Innovation (STI) policies including other important components, such as the promotion of South–South and North–South–South cooperation, investments at country level into accessible and quality education systems, gender equality, the reinforcement of science–policy–society interfaces and the inclusion of national, regional, and grassroots innovation capacities, as well as of local and indigenous knowledge.* Today, more than ever, we need more science, *better, interconnected, crosscutting science, relevant to people.*

The adoption of the United Nations' 2030 Agenda in September 2015, with its set of 17 Sustainable Development Goals (SDGs), marked a significant step forward in the recognition of the contribution of STI to sustainable development and its three pillars: economic, social, and environmental. The 2030 Agenda also offers immense opportunities to reconnect science to society and to build a new basis for research and development as a key precondition for both science and society to flourish.

As the only UN agency with science in its mandate, UNESCO has a leading role in using and promoting STI as effective tools to contribute to sustainable development. Since its foundation 72 years ago, the Organization has been strongly committed to reinforcing the links between science, policy, and society, and to promoting STI policies that benefit society as a whole. With its network of international scientific programmes, centers of excellence, institutes, and Chairs worldwide, UNESCO has an important role to play in the common effort to achieve the SDGs.

This publication allows a larger audience to benefit from the high-level researches presented during the 2016 edition of the Conference, in key areas for sustainable development such as energy, disasters risk reduction, medical technologies, urban development, ICT, and humanitarian action.

EPFL is hosting the UNESCO Chair in Technologies for Development since 2007. UNESCO is grateful for its remarkable work in collaboration with partners from emerging and developing countries, which contributes to poverty reduction and sustainable development.

Geneva, Switzerland
2017

Flavia Schlegel United Nation, Education, Scientific and Cultural Organization—UNESCO

Acknowledgements

The editors would like to thank the many individuals and organizations who generously contributed their time, insight, and support. We would like to begin with the members of the Scientific Committee and our Session Leaders, who guided the conference preparation.

We would also like to express our thanks to Prof. Philippe Gillet, Vice-President of Academic Affairs at the Ecole Polytechnique Fédérale de Lausanne (EPFL), for his presence at the Conference and unfailing support of the Cooperation & Development Center (CODEV).

Through openly sharing their great expertise and diverse perspectives, the speakers at the UNESCO Conference substantially contributed to its success. Our heartfelt thanks go to Dr. Flavia Schlegel (UNESCO), Dr. María Fernanda Espinosa (United Nations Office), Ms. Barbara Bulc (Global Development), Mr. Yves Daccord (ICRC) and Dr. Ashok Gadgil (University of California, Berkely) for their highly appreciated involvement and support.

This project likewise could not have succeeded without the quality and diversity of the various authors' and researchers' contributions. In response to the call for papers, the Scientific Committee evaluated over 156 papers and ultimately selected 125 to be presented at the Conference. Of these, 17 finalists were chosen based on the following criteria: (1) innovative concept and research questions; (2) potential social impact of the application; (3) contribution to the discipline as whole; and (4) clarity and understandability. We express our appreciation to all of these authors, without who this publication would not have been possible.

In addition, we would like to very warmly thank Mr. Emmanuel Estoppey and his team from the Lavaux UNESCO World Heritage Site, who went out of their way to welcome us for the social event in the charming village of Grandvaux.

Our sincere thanks also go to the Ingénieurs du Monde (IDM) team and our colleagues at CODEV, who contributed extensively to the organization of this conference.

Finally, we are very grateful for the generous patronage of the Swiss Agency for Development and Cooperation (SDC), the Canton de Vaud, the City of Lausanne, the Swiss National Science Foundation (SNSF), the ICRC, and the conference sponsors. Their support and their partnership are critical for bringing us to reflect on how technological innovation can lead to stronger social impact and lead the way toward more suitable development at global level.

Scientific Committee and Session Leaders

Pankaj Agarwal, Panitek AG, Liechtenstein
Bipasha Baruah, Western University, Canada
Justin Bishop, University of Cambridge, UK
Jennifer Brant, Innovation Insights, Switzerland
Leo Anthony Celi, Harvard Medical School, Beth Israel Deaconess Medical Center, and Massachusetts Institute of Technology, USA
Albrecht Ehrensperger, University of Bern, Switzerland
Marie-Valentine Florin, International Risk Governance Council, Switzerland
Zach Friedman, LIGTT: Institute for Globally Transformative Technologies, USA
Ashok J. Gadgil, University of California, Berkeley, USA
Mini Govindan, The Energy and Resources Institute, India
Sachiko Hirosue, Ecole Polytechnique Fédérale de Lausanne, Switzerland
Silvia Hostettler, Ecole Polytechnique Fédérale de Lausanne, Switzerland
Tunde Kallai, PASRI—ANPR, Meta Group, TR-Associates Ltd, Switzerland
Prabhu Kandachar, Delft University of Technology, The Netherlands
Walter Karlen, ETH Zurich, Switzerland
Denisa Kera, University of Singapore, Singapore
Bertrand Klaiber, Ecole Polytechnique Fédérale de Lausanne, Switzerland
Papa Amadou Konte, Dakar City Municipality, Senegal
Paula Lytle, The World Bank, United States
Temina Madon, University of California, Berkeley, USA
Charles Martin-Shields, George Mason University, USA
Kinsuk Mitra, InsPIRE Network for Environment, India
François Münger, Swiss Agency for Development and Cooperation, Switzerland
Hung Nguyen-Viet, Hanoi School of Public Health, Vietnam
Vipan Nikore, Cleveland Clinic and Massachusetts Institute of Technology, USA
Ermanno Pietrosemoli, Abdus Salam International Centre for Theoretical Physics, Italy and Fundación "EsLaRed," Venezuela
María Catalina Ramírez, Universidad de los Andes, Colombia
Federico Rosei, University of Quebec, Canada
Hans Schaffers, Aalto University, Finland

Klaus Schönenberger, Ecole Polytechnique Fédérale de Lausanne, Switzerland
Tobias Siegried, hydrosolutions Ltd., Switzerland
Lucy Stevens, Practical Action, UK
Andrés Felipe Valderrama Pineda, Aalborg University, Denmark
Christian Zurbrügg, Eawag: Swiss Federal Institute of Aquatic Science and Technology

Contents

Editors and Contributors

About the Editors

Silvia Hostettler is an environmental scientist and international development scholar with 19 years of experience in researching sustainable development challenges in the global South. She is interested in how technological innovation can support poverty reduction and sustainable development. She studied at the University of Geneva, the University of Aberdeen (UK) and at EPFL, conducted research in Burkina Faso, Ghana and Mexico and was based in Bangalore for four years as Executive Director of swissnex India. Currently, she is the Deputy Director of the Cooperation and Development Center at EPFL and directs the International Conference of the UNESCO Chair in Technologies for Development, e-mail: silvia.hostettelr@epfl.ch.

Samira Najih Besson is an engineering geologist with option in hydrogeology from Lausanne University. She studied at the University of Geneva and the Cadi Ayyad University of Marrakech. She has worked for Swiss public institution and international humanitarian organizations where she conducted projects on GIS, environmental studies, natural hazard disaster management. Currently, she is the project manager of the International Conference of the UNESCO Chair in Technologies for Development at CODEV at EPFL, e-mail: samira.najih@epfl.ch.

Jean-Claude Bolay is the Director at CODEV. As the scientific director of the Center, he leads a team of 25 scientists and collaborators with the goal of coordinating development cooperation activities at EPFL. Among many training, research, and management activities, the Center manages the UNESCO Chair in Technologies for Development, which focuses on the adaptation of technologies and innovation for the global South, e-mail: jean-claude.bolay@epfl.ch.

Contributors

Maxime Audouin Chair Management of Network Industries (MIR), EPFL, Lausanne, Switzerland

Elizabeth Bailey Consortium for Affordable Medical Technologies, Boston, MA, USA

Alejo Cochachin Rapre Unidad de Glaciología y Recursos Hídricos (UGRH), Autoridad Nacional de Agua, Huaraz, Peru

Katharine G. Broach Hutchins Consortium for Affordable Medical Technologies, Boston, MA, USA

Cristina Coscia Department of Architecture and Design, Politecnico di Torino, Turin, Italy

Malika Davids University of Cape Town, Cape Town, South Africa

Rahel Dette The Berlin-Based Global Public Policy Institute (GPPi), Berlin, Germany

Keertan Dheda University of Cape Town, Cape Town, South Africa

Sunil Dhingra The Energy and Resources Institute (TERI), New Delhi, India

Nora Engel Maastricht University, Maastricht, Netherlands

Emily Eros American Red Cross, Washington, DC, USA

Flavia Farina Basin Modeling Laboratory, Institute of Geosciences, Federal University of Rio Grande do Sul, Porto Alegre, Brazil

Francesca De Filippi Department of Architecture and Design, Politecnico di Torino, Turin, Italy

Matthias Finger Chair Management of Network Industries (MIR), EPFL, Lausanne, Switzerland

Javier Fluixá-Sanmartín Centre de Recherche sur l'Environnement Alpin (CREALP), Sion, Switzerland

Holger Frey Department of Geography, University of Zurich (UZH), Zurich, Switzerland

Javier García Hernández Centre de Recherche sur l'Environnement Alpin (CREALP), Sion, Switzerland

Joshua Goldsmith University of Geneva, InZone, Geneva, Switzerland

César Alfredo Gonzales Alfaro CARE Perú, Huaraz, Peru

Darelle van Greunen The Centre for Community Technologies, Nelson Mandela University, Port Elizabeth, South Africa

Smitha Gudapakkam Consortium for Affordable Medical Technologies, Boston, MA, USA

Roberta Guido Department of Architecture Design and Urban Planning, University of Sassari, Sassari, Italy

Erin Hayba University of Geneva, InZone, Geneva, Switzerland

Ulrik Birk Henriksen Technical University of Denmark (DTU), Copenhagen, Denmark

Mark Herringer Healthsites.io, London, UK

Silvia Hostettler Cooperation and Development Center (CODEV), Ecole Polytechnique Fédérale de Lausanne (EPFL), Lausanne, Switzerland

Christian Huggel Department of Geography, University of Zurich (UZH), Zurich, Switzerland

Pierre Jaboyedoff Effin Art, Lausanne, Switzerland

Sashidhar Jonnalagedda Program Essential Tech Cooperation and Development Center EPFL, Lausanne, Switzerland

Dikolela Kalubi International Committee of the Red-Cross, Lausanne, Switzerland

David R. King Harvard Medical School, Boston, USA; Massachusetts General Hospital Department of Surgery, Boston, USA

Raj Madhavan Humanitarian Robotics and Automation Technologies, Clarksburg, USA

Bhanu Mall Poorvanchal Gramin Vikas Sansthan (PGVS) Organization, Lucknow, India

Paul Andree Masías Chacón Corporación RD S.R.L, Cusco, Peru

Aikaterini Mantzavinou Harvard-MIT Program in Health Sciences and Technology, Cambridge, MA, USA

Tamara Maričić Institute of Architecture and Urban & Spatial Planning of Serbia, Belgrade, Serbia

Sally Miller MIT Department of Mechanical Engineering, Cambridge, USA

Sanghamitra Misra Oxfam, New Delhi, India

Robert Monné Utrecht University, Utrecht, Netherlands

Barbara Moser-Mercer University of Geneva, InZone, Geneva, Switzerland

Dennis Nagle MIT D-Lab, Cambridge, USA

Kristian R. Olson Consortium for Affordable Medical Technologies, Boston, MA, USA

Madhukar Pai McGill University, Montreal, Canada

Nitika Pant Pai McGill University, Montreal, Canada

Edson Prestes Phi Robotics, Institute of Informatics, Federal University of Rio Grande do Sul, Porto Alegre, Brazil

Bryan J. Ranger Harvard-MIT Program in Health Sciences and Technology, Cambridge, MA, USA

Leonardo Renner Basin Modeling Laboratory, Institute of Geosciences, Federal University of Rio Grande do Sul, Porto Alegre, Brazil

Luis Meza Román Municipalidad de Carhuaz, Carhuaz, Peru

Eric de Roodenbeke International Hospital Federation, Bernex, Switzerland

René Saameli International Committee of the Red-Cross, Lausanne, Switzerland

Yeeshu Shukla Christian Aid Organization, London, UK

Tatiana Silva Basin Modeling Laboratory, Institute of Geosciences, Federal University of Rio Grande do Sul, Porto Alegre, Brazil

Shirish Sinha Swiss Agency for Development and Cooperation (SDC), New Delhi, India

Robert Smalley Harvard Medical School, Boston, USA

Marco René Spruit Utrecht University, Utrecht, Netherlands

Tim Sutton Healthsites.io, London, UK

Barkha Tanvir The Energy and Resources Institute (TERI), New Delhi, India

Debbie L. Teodorescu MIT D-Lab, Cambridge, USA

Alida Veldsman The Centre for Community Technologies, Nelson Mandela University, Port Elizabeth, South Africa

Koneru Vijaya Lakshmi Society for Development Alternatives, New Delhi, India

Miodrag Vujošević Institute of Architecture and Urban & Spatial Planning of Serbia, Belgrade, Serbia

Rodrigo Wiebbelling Basin Modeling Laboratory, Institute of Geosciences, Federal University of Rio Grande do Sul, Porto Alegre, Brazil

Vijayashree Yellappa Institute of Public Health, Bangalore, India

Slavka Zeković Institute of Architecture and Urban & Spatial Planning of Serbia, Belgrade, Serbia

Daniel Ziegerer Swiss Agency for Development and Cooperation (SDC), New Delhi, India

Marc Jan Christiaan van den Homberg Cordaid and TNO, The Hague, Netherlands

Part I
Introduction

Chapter 1
From Innovation to Social Impact

Silvia Hostettler

1.1 What Is Innovation?

Today, there appears to be a widespread call for innovation: product innovation, process innovation, market innovation, organizational innovation, and social innovation. It sometimes feels as though, when at loss, we call upon innovation. The origin of the word *innovation* means "restoration, renewal," from the Latin *innovationem* and *innovare*. Joseph Schumpeter is considered as the first economist to have drawn attention to the importance of innovation in the 1930s (Croitoru 2012; Schumpeter 1911). Innovation can be a new method, idea, or product—something that is new or different. Innovation's key characteristic is that it is assumed to provide a significant, positive change. "*To be called an innovation, an idea must be replicable at an economical cost and must satisfy a specific need* [...]".[1] In the context of development, we look more specifically at social innovation, which can be described as "... *a novel solution to a social problem that is more effective, efficient, sustainable, or just than current solutions. The value created accrues primarily to society rather than to private individuals.*"[2]

[1]http://www.businessdictionary.com/definition/innovation.html.

[2]https://www.gsb.stanford.edu/faculty-research/centers-initiatives/csi/defining-social-innovation.

S. Hostettler (✉)
Cooperation and Development Center (CODEV), Ecole Polytechnique Fédérale de Lausanne (EPFL), Lausanne, Switzerland
e-mail: silvia.hostettler@epfl.ch

S. Hostettler et al. (eds.), *Technologies for Development*,
https://doi.org/10.1007/978-3-319-91068-0_1

Innovation is often divided into two broad categories: evolutionary innovation (also called continuous or dynamic evolutionary innovation) brought about by many incremental advances in technology or processes, and revolutionary innovation (also called discontinuous innovation) which is often disruptive and new.[3] In the technologies for development field, both types of innovation exist. When considering the significant needs that continue to go unmet in the Global South, frugal innovation is particularly important. Frugal innovation is a process whereby new business models are developed, value chains are reconfigured, and products are redesigned in a scalable, sustainable manner to serve users facing extreme affordability constraints: "*Simple, frugal innovation provides functional solutions using scant resources for the many who have little means*".[4] Frugal innovation implies doing better with less by focusing on affordability, simplicity, quality, and sustainability.[5]

1.2 Progress Driven by Technological Innovation

The innovation that interests us here is technological innovation for sustainable development in the Global South. Since 1990, a billion people have escaped extreme poverty, 2.1 billion have gained access to improved sanitation, and more than 2.6 billion have gained access to an improved source of drinking water (United Nations 2015). Between 1990 and 2015, the global under-five mortality rate drastically decreased from 91 per 1000 live births to 43. Between 2000 and 2015, the incidence of HIV, malaria, and tuberculosis declined. The proportion of seats held by women in parliaments worldwide, though still a far cry from egalitarian representation, rose from 17 to 23% in 2016 (UNDP 2016). Technological innovation has played an important role in this progress. New technologies can help governments and citizens to interact more efficiently and increase the scope and efficiency of public services. With the steadily growing penetration rate of mobile phones, many countries are now able to use mobile phones to extend basic social services, including health care, financial services and education, to hard-to-reach populations. The Internet allows for considerably more information sharing than any other means of communication ever has (UNDP 2016).

Technology is one of the key factors that can help developing countries close the gap with industrialized countries. In addition to infrastructure, a productive and healthy workforce, roads, and access to information and knowledge, technology can help countries to leapfrog forward (Sachs 2015; Wooldridge 2010). For instance, now that traditional sources of energy such as fossil fuels are coming to an end, adopting renewable energies based on hydro-, wind-, or solar power might offer an opportunity for developing countries to not repeat the same mistakes made by industrialized countries, but instead forge ahead with the help of cutting-edge

[3]http://www.businessdictionary.com/definition/innovation.html.

[4]http://www.frugal-innovation.com/what-is-frugal-innovation.

[5]https://hbr.org/2014/11/4-ceos-who-are-making-frugal-innovation-work.

technology in the energy and health sectors. Mobile technology can be transformative. For instance, mhealth apps can help upscale health programmes for prenatal care. It can also raise awareness about the risk of contracting malaria and smart phones can now diagnose pneumonia via diagnostic devices (Ettinger et al. 2016; Friedman and Karlen 2015). Mobile technology can also leverage social impact in the financial sector by providing remote banking services for rural low-income communities and information for farmers about fair market prices (Martin and Abbot 2011).

1.3 Remaining Challenges

However, as outlined in the 2030 Agenda for Sustainable Development, many needs are yet to be met in the Global South.[6] Human development has been uneven, as progress has bypassed many communities; others have merely managed to ensure basic human needs. Even though poverty has been reduced massively over the past 25 years, poor nutrition still causes 45% of the deaths among children under five. Stunting and other delays in physical development are still very common in children in developing countries. Yet, a third of the world's food supply is wasted each year. By reducing this figure to 25%, 870 million more people could be fed. Unless the deprivation is addressed, 167 million children will live in extreme poverty by 2030, and 69 million children under five will die of preventable causes. These outcomes will undoubtedly have a negative impact on the capacities of future generations. 114 million young people and 644 million adults still lack basic reading and writing skills. Persistent deprivation is observable in various aspects of human development. Yet, the income gap continued to widen in 34 of the 83 countries observed between 2008 and 2013. In 23 countries, the poorest 40% saw their income decline; and yet, alarmingly, income growth has been particularly pronounced at the top rungs of the income ladder—in other words, the rich get richer while the poor get poorer. Approximately 46% of the total increase in income between 1988 and 2011 was attributed to wealthiest 10% of the population. Since 2000, 50% of the increase in global wealth benefited only the wealthiest 1% of the world's population; only 1% went to the poorest 50%. Global wealth has become far more concentrated. In 2000, the wealthiest 1% of the population held 32% of global wealth. This increased to 46% in 2010 (UNDP 2016). Not surprisingly, new development challenges have emerged and/or deepened, including climate changes, conflict, and desperate migration (UNDP 2016).

[6]http://www.un.org/sustainabledevelopment/development-agenda/.

1.4 Need for Social Impact

Innovation is the new buzzword. Much hope has been placed on technological innovation, social innovation, financial innovation, and organizational innovation; we might even start hearing talk about the need to *innovate innovation*. However, innovation in itself is not enough. Social impact—meaning positive change for society and, in this case, low-income communities in the Global South—requires successful implementation and use of technologies at a large scale. Why, with rampant technological innovation, does the social impact of technology remain so limited? There are still many needs unmet in many parts of the world, and much hope is being placed on innovation to accelerate the implementation of the SDG's that aim for significant social impact. This chapter explores some of the decisive key factors when considering how we can move from innovation to social impact.

Figure 1.1 shows a thermal water heating system based on photovoltaic pipes in Ghandruk, Nepal. Initial cost of USD 400 for an average use life of 10 years. Enough hot water is provided for six hot showers for tourists per day, providing an additional sustainable opportunity for revenue creation.

Fig. 1.1 Thermal water heating system

1.5 The Bumpy Road to Social Impact

Developing successful technologies is challenging, and obtaining the desired social impact is even more so. For an innovation to have social impact, it must make the transition from an innovation to a technology that can be implemented at scale, e.g., by becoming a mainstream product such as a smart phone. The key question is how can a technology be brought to scale in order to have a broad and positive impact? Ensuring that a technological solution successfully addresses a specific issue in the Global South requires careful attention during each phase of production, from the initial idea to bringing the technology to scale (Hostettler 2015). The first step is conducting a thorough needs assessment in order to ensure that intended beneficiaries' priority needs are being targeted and that the right population has been identified. The needs assessment can also help indicate whether a technology will be socioculturally appropriate and therefore increase the chances of adoption. During prototype development in collaboration with key stakeholders, the cost of the technology—a crucial factor—must be carefully considered, as well as the customers' ability and willingness to pay for it. An unaffordable technology will not have the desired impact, as no one will be able to purchase it. Developing a sustainable business model from the outset of technological innovation is key. If economic insight is not part of the technological innovation process, then the chances of failure increase dramatically, as the long-term financial sustainability cannot be ensured.

Regarding technical aspects, especially in developing countries, a technology's robustness is crucial, as it will have to withstand high temperatures, humidity, dust, and unstable electrical circuits. Other key questions to be considered are: Is the technology easy to use, or can it only be operated by experts? In the event of a breakdown, is there a supply chain for spare parts, or will they need to be imported at a prohibitive cost from developed countries? Do the required capacities and infrastructure for repairs exist? Can the waste products of production and the technological product be recycled at the end of its life? Does the product meet the objectives of a circular economy? How can local staff be trained? Does the technology comply with national and international standards? What needs should be considered regarding the legal framework, e.g., patenting or open access? In addition, we need to ask ourselves, does this technology push local companies out of business? Does it consider local political factors such as corruption and civil unrest? Who will have access to this technology? Does it run the risk of creating inequalities?

It is particularly important to integrate scalability early on. Key factors such as country size, the political landscape, culture, language, the potential cost of establishing a regional service network, logistical challenges, spare part depots, and human resources must be taken into account. A large-scale study of 20,000 geotagged households in Kenya showed that 50% of unconnected homes are "under grid," meaning they are within range of an existing transformer but are not connected. It turned out that the need for innovation lies not at the technological level but in identifying appropriate tech adoption incentives, such as subsidies and innovative financing mechanisms (Lee et al. 2016). This experience shows the importance of taking an interdisciplinary approach by bringing together practitioners, engineers, anthropologists, economists, computer scientists etc. to develop innovation that can bring about a large-scale, positive social impact.

The constraints of low-resource settings can be a strong driver of innovation. People living in rural or urban areas, educated or uneducated, are not just consumers of innovation, they can also be the source. They have significant inventive power to design and solve problems locally in their own sociocultural context. In Togo, for instance, a 3D-printer was made using electronic waste salvaged from landfill sites, with the aim of improving the lives of communities by "printing" objects such as medical prostheses.[7] 3D printing technology has the potential to bring about concrete social impact and is particularly interesting in emergency relief situations and for providing medical services in remote areas. Medical devices can be printed on demand and in adequate quantity, e.g., irrigation syringes, oxygen splinters, umbilical cords, and prostheses. Furthermore, using portable solar-powered 3D printers can increase the technology's autonomy. Overall, 3D printing could not only improve health care in the developing world but could also allow for economic independence. 3D printing might help countries launch their own production rather than depend on global supply chains by importing expensive medical devices from the developed world (Dotz 2015).

1.6 Conclusion

Social impact requires the successful implementation of a technology at a large scale. In this respect, developing a sustainable business model is crucial. De Jaeger et al. (2017) argue that the underlying challenge when it comes to ensuring a high impact depends on the development and successful implementation of a robust innovation/entrepreneurship ecosystem, the cooperation of all stakeholders, and sufficient resources. Innovation must also be linked to national and institutional systems, e.g., when developing medical technologies, the Ministry of health should be a key partner from the onset.

The challenge consists in considering all of these factors simultaneously; by not doing so, the entire arch of technology development could run the risk of collapse due to a single factor (e.g., financial sustainability or socio-cultural acceptability) not being adequately addressed. It is for this reason that universities have a key role to play in supporting the path to innovation and entrepreneurship education, with outcomes that impact society at large. Human-centered design is the core of development engineering that aims to scale for impact by incorporating development goals, constraints, and opportunities (Levine et al. 2016). Development engineering is based on the belief that innovative technologies have the potential to improve life in low-income communities by incorporating insight from the social sciences throughout technological innovation—from prototyping to production at scale.

[7] http://observers.france24.com/en/20161110-togolese-invent-3d-printer-waste.

We increasingly realize that engineers not only need to excel in their discipline but must also be able to work efficiently in economically, socially, and environmentally diverse contexts. To create a new generation of practitioners and social entrepreneurs, formal training at academic institutions must strive to include additional skill sets based on interdisciplinary training and design-based thinking in order to bridge the gap between innovation and social impact.

References

Croitoru, A. (2012). A review to a book that is 100 years old. *Journal of Comparative Research in Anthropology and Sociology, 3*(2), 137–148.

De Jager, H. J., et al. (2017). Towards an innovation and entrepreneurship ecosystem: A case study of the Central University of Technology, Free State. *Science Technology & Society, 22*(2), 310–331.

Dotz, D. (2015). A Pilot of 3D printing of medical devices in Haiti. In S. Hostettler, E. Hazboun, & J.-C. Bolay (Eds.), *Technologies for development: what is essential?* (pp. 33–44). Paris: Springer.

Ettinger, K. M., Pharaoh, H., Buckman, R. Y., Conradie, H., & Karlen, W. (2016). Building quality mhealth for low resource settings. *Journal of Medical Engineering & Technology, 40*(7–8), 431–443.

Friedman, Z, & Karlen, W. (2015). Medical devices and information communication technologies for the base of the pyramid. In S. Hostettler, E. Hazboun & J. -C. Bolay (Eds.), *Technologies for development: What is essential?* (pp. 113–118). Berlin: Springer.

Hostettler, S. (2015). Technologies for development. What really matters? In S. Hostettler, E. Hazboun, & J. -C. Bolay (Eds.), *Technologies for development: What is essential?* Paris: Springer.

Lee, K., Brewer, E., Christiano, C., Meyo, F., Miguel, E., Podolsky, M., et al. (2016). Electrification for 'under grid' households in rural kenya. *Development Engineering, 1,* 26–35. https://doi.org/10.1016/j.deveng.2015.12.001.

Levine, D. I., Lesniewski, M. A., & Agogino, A. M. (2016). Design thinking in development engineering. *International Journal of Engineering Education, 32*(3B), 1396–1406.

Martin, Brandie Lee, & Abbott, Eric. (2011). Mobile phones and rural livelihoods: Diffusion, uses, and perceived impacts among farmers in rural Uganda. *Information Technologies & International Development, 7*(4), 17–34.

Sachs, J. (2015). *The age of sustainable development* (p. 521). New York: Columbia University Press.

Schumpeter, J. A. (1911, 2008). *The theory of economic development: an inquiry into profits, capital, credit, interest and the business cycle* (R. Opie, Trans.). New Brunswick, U.S.: Transaction Publishers.

United Nations. (2015). *The millennium development goals report 2015.* http://www.un.org/millenniumgoals/2015_MDG_Report/pdf/MDG%202015%20Summary%20web_english.pdf Accessed November 1, 2016.

United Nations Development Programme (UNDP). (2016). *Human development report 2016: Human development for everyone.* Retrieved from http://hdr.undp.org/sites/default/files/2016_human_development_report.pdf.

Wooldridge, A. (2010, April 17). The world turned upside down. A special report on innovation in emerging markets. *The Economist*, pp. 1–14.

Part II
Humanitarian Technologies

Chapter 2
Do No Digital Harm: Mitigating Technology Risks in Humanitarian Contexts

Rahel Dette

2.1 Introduction

> … which reminds us of the fact that peripheral populations are being subjected to more or less experimental technologies.
>
> Katja Lindskov Jacobsen, The Politics of Humanitarian Technology (2015)

In humanitarian emergencies, information communication technologies (ICTs) such as mobile phones and web-based platforms offer powerful tools for communicating with communities, remote needs assessments and data collection. A quickly growing literature confirms the benefits that ICTs can offer to aid efforts with regards to efficiency, effectiveness and accountability (Raftree and Bamberger 2014; Kalas and Spurk 2011). These promises are especially pronounced in insecure environments, where access constraints hinder aid actors from reaching local populations, such that digital channels could be the only way to send and receive critical information. At the same time, the potential consequences of implementing technology-based projects poorly or overseeing unintended consequences can be detrimental and sometimes lethal.

Recognizing the challenges and risks with technologies can help avoid pitfalls and unintended digital harm. ICTs are known to introduce complications in a number of ways: Digital tools themselves alter the interaction between aid staff and recipients, which can add to and exacerbate crises or conflict dynamics (Jacobsen 2015; Vazquez and Wall 2014; Altay and Labonte 2014). The digitization of communications introduces new security and privacy risks as data transmitted on electronic devices or networks becomes susceptible to third party interception and breaches, sometimes unnoticeably (Internews 2015; Schneier 2015). However, such challenges entangled with using ICT for humanitarian purposes have not been adequately researched or addressed in literature and practice. Although concrete proposals for new ethics and

R. Dette (✉)
The Berlin-Based Global Public Policy Institute (GPPi), Berlin, Germany
e-mail: rdette@gppi.net

S. Hostettler et al. (eds.), *Technologies for Development*,
https://doi.org/10.1007/978-3-319-91068-0_2

conventions to guide technology uses are increasingly considered necessary, they are holding off (Raymond and Card 2015a; Gilman and Baker 2014; Sandvik et al. 2014). While a number of promising initiatives develop guidelines and good practice lessons, they tend to focus on disaster settings, explicitly omit recommendations for complex, man-made emergencies (GSMA 2012; UAViators 2015; Madianou et al. 2015). In conflict zones, aid actors are left to make up rules as they go, or forfeit opportunities by opting against technologies altogether, but these decisions are typically not documented (Raymond et al. 2013; Steets et al. 2015). A better understanding of the perceived and real risks entangled in the use of ICT, as outlined here, can help inform a responsible, sustainable humanitarian technology practice that works in all settings.

This paper draws on finding of the 3-year research project 'Secure Access in Volatile Environments (SAVE)' that was undertaken by Humanitarian Outcomes (HO) and the Global Public Policy Institute (GPPi) with funding from the UK Department for International Development (DFID). Part of the research assessed technologies that aid actors can use for monitoring and evaluation (M&E) in insecure and hard-to-reach areas. The research was undertaken in close collaboration with NGOs and UN agencies in Afghanistan, Somalia, South Sudan and the Syria region with the aim to provide practical contributions. A 'menu of technology options for monitoring' introduces and explains select ICT tools in detail, zooming in on challenges in order to propose effective risk mitigation strategies. The menu and this paper recognize the significant potential technologies offer to humanitarian efforts, but caution against rushed implementation at the risk of overseeing severe challenges and limitations. A risk-aware approach to all new tools, especially those that are digital or data-based, can help aid actors assure they 'do no digital harm.'

2.2 Technology Advantages

ICT in insecure environments

ICT aids communication and can be used in aid efforts. It includes mobile phone, location trackers, software, web-based platforms, digital media and more. Here, we focus on those tools that are already widespread or easily accessible in crisis contexts and thus ready and reliable to use even in precarious situations. The narrow focus on insecurity and conflict settings, where information is often sensitive, makes risks and challenges especially apparent, and can help inform mitigation and best practices that translate to other settings.

In insecure environments, several limitations curtail the selection of ICTs that can be used. Access restrictions, poor infrastructure, budgets constraints and high levels of uncertainty require tools to function without constant electricity supply, across wide distances and without advanced IT support. In close collaboration with aid organizations and technology experts, we identified four technology types that meet these criteria: mobile phone-based feedback mechanisms, handheld devices

Table 2.1 Technological tools for remote monitoring and communication in insecure environments: Types, uses, and challenges

Phone-based feedback mechanisms and two-way communication	Digital data collection with smartphones and tablet computers	Remote sensing with satellites or UAVs and location tracking (GPS)	Broadcasting and community production of radio shows
Where basic mobile phones are widely spread, they offer reliable channels to reach local communities. Calls, text messages and interactive voice recordings (IVR) can be processed by call centres or specialized software	Aid organizations can use digital data entry linked to electronic databases to replace paper-based survey instruments and create faster, more automatic data analysis. Small handhelds are also more unobtrusive than clipboards	High-resolution geospatial imagery analysed by experts can elucidate context conditions, observable changes and outcomes of interventions, as well as population movement. Radars and sensors can capture unique data	Broadcast radio can spread humanitarian information, conflict or aid delivery updates. To target information, aid staff can stream pre-recorded shows in select locations. Interactive radio shows station can also receive feedback
Complaints and information hotlines Phone-based household surveys Verification calls Focal point reports	Surveys and questionnaires Registration and distribution reporting GPS- and timestamps in surveys	Observation and analysis with satellite UAV imagery for close-up analysis Radar/sensor data Barcode tracking	Outreach, advocacy and engagement Publicize/explain feedback channels Community radio
⊗ Proliferation of parallel hotlines can lead to confusion ⊗ Security and privacy risks to staff and aid recipients ⊗ Risk of bias towards those owning phones (often men) ⊗ Volatility due to poor infrastructure	⊗ Devices can raise the visibility of aid staff and mistrust among authorities and locals ⊗ Competing tools can cause fragmentation ⊗ Privacy/security risks digitizing data: theft, interception, surveillance, etc.	⊗ Lack of guidance/established practice ⊗ Can expose vulnerable groups ⊗ High costs can deter organizations as effect and return are not always clear ⊗ Dependency on image providers, data brokers and experts	⊗ One-way radio broadcasts cannot record feedback ⊗ Difficult to measure impact and identify the audience ⊗ Security risks due to high visibility: interception possible ⊗ Gender bias towards male voices

for digital data collection, remote sensing with satellites or unmanned aerial vehicles (UAVs) and broadcasting with radios. They enable a range of functions to complement communication efforts but also introduce new challenges as summarized in Table 2.1

ICTs in humanitarian action

Aid organizations report a number of advantages with technologies, especially around saving costs and time. In highly insecure settings, ICT increasingly facilitates direct two-way interactions that otherwise could not take place. Where aid access is less restricted, many find that face-to-face time with local communities can be used more efficiently when survey data can be entered directly into digital devices. Because these

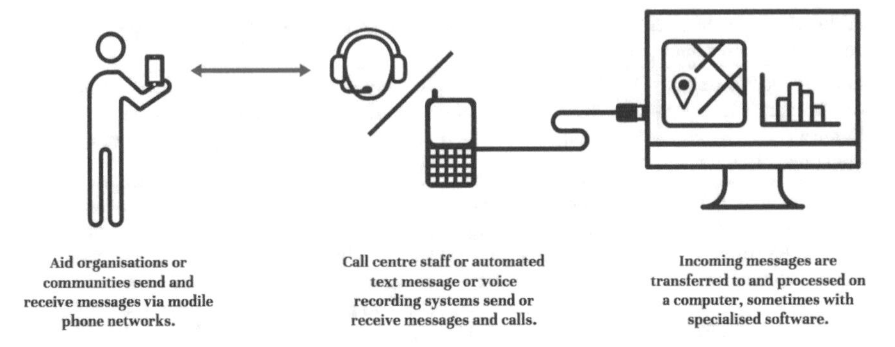

Fig. 2.1 Technology Type 1—Phone-based feedback mechanisms

typically process and upload information directly to databases, the turnaround for data analysis and use increases significantly. ICTs were often praised to be convenient, customizable and very good at handling data efficiently. The devices themselves can be unobtrusive or unnoticeable, which sometimes improved the security for aid staff and recipients. A helpful way to conceptualize such benefits as well as associated shortcomings is what one researcher coined the 'law of amplification' (Toyama 2015). Technology can amplify the intent and capacity of stakeholders, he says, but never substitute for deficiencies. A closer look at how this plays out for different technologies helps lay the ground for afterwards deciphering how problems, too, are amplified—or newly introduced (Fig. 2.1).

Countless studies cite and investigate the 'unprecedented,' 'ubiquitous' spread of basic mobile phone worldwide and the impact their availability and affordability can have in development and humanitarian contexts (Hallow et al. 2012; de Montjoye et al. 2014). Calls or SMS are comparatively cheap and very quick and, importantly, often come naturally to local communities in crisis-affected areas. This makes mobile phones ideal for reaching out to and being reached by more people (Robinson and Obrecht 2016). Aid actors have started integrating phones proactively by offering hotlines for aid recipients, by calling households to collect data, and by inviting comments, complaints or suggestions via SMS (Korenblum 2012). More and more 'hotlines' also are provided via WhatsApp and other online messaging tools that work on smartphones as well as desktops. Powered by mobile data or Internet connection and offering encrypted channels, these started replacing SMS and calls in areas where smartphones are spreading, for example, in Syria. Aid agencies also make use of phones for distributing mobile vouchers and cash directly to people's mobiles or to quickly spread messages with warnings, updates or important information. While regular visits to communities take weeks, SMS updates or survey questions sent to phones take hours or minutes to share and process (Fig. 2.2).

Summing up the significant improvements digital surveys offers data collection, one aid worker said: 'There are no more pregnant men anymore.' Supervisors can programme questionnaires to require specific answers to specific questions, and skip irrelevant questions, such as asking men whether they are pregnant. Spell check and

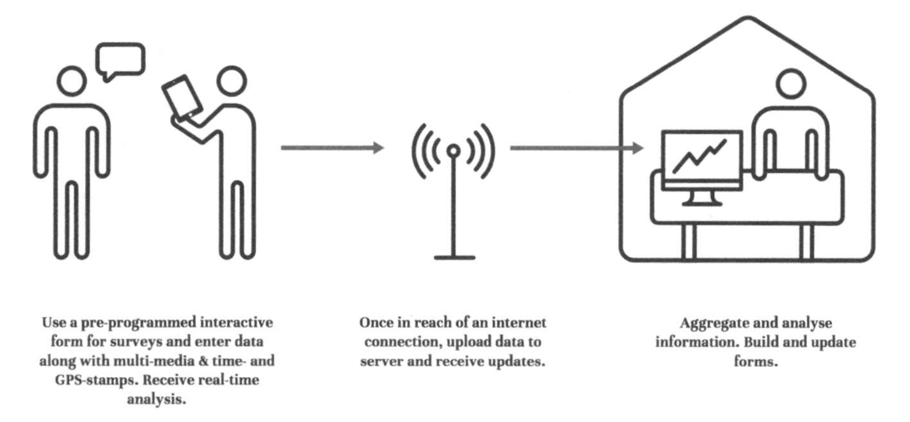

Fig. 2.2 Technology Type 2—Digital data collection

error detection also improve data entries, thereby preventing accidental nonsensical answers. Altogether, running surveys on digital devices rather than paper was largely seen to speed up interviews, sometimes cutting their time in half. Where an Internet connection is available, data transfer from the field or field office to headquarters can be immediate, and makes laborious manual data entry and transfer unnecessary, thus cutting weeks' worth or time down to hours. 'Reporting used to be a headache,' one monitoring officer said. 'With smartphones, it no longer is.' During interviews, smartphones sometimes offered unexpected further benefits, like drawing less attention from members of the community than a flipboard would and sometimes giving survey respondents more in-the-moment privacy by typing rather than saying answers. In addition, field office staff reported that GPS- and timestamps provide certainty that interviews are actually completed at different times with community respondents in different locations. In Somalia, for example, enumerators were suspected to sometimes fill out forms in bulk, inventing answers. Typically, this would happen when it was too difficult or dangerous to reach aid recipients. Because the digital data entries made this trend apparent, aid organizations could address this issue and start constructive conversations with enumerators when and where data collection was too dangerous (Fig. 2.3).

Aerial imagery and geospatial analysis can capture independent and objective information from areas that are too remote or insecure to reach or where larger patterns may not be observable up-close. Where access is restricted, this data can provide valuable insights on infrastructure and shelter, vehicle positions and the effects of disasters including flooding, drought or landslides. Captured repeatedly over time, imagery can help assess project outcomes and, in some contexts such as agricultural intervention, impact. Remote sensing or 'earth observation' information is often visualized on maps or triangulated with other data sets. This is especially beneficial for making sense of complex datasets and putting information in context (Fig. 2.4).

Fig. 2.3 Technology Type 3—Remote sensing and location tracking

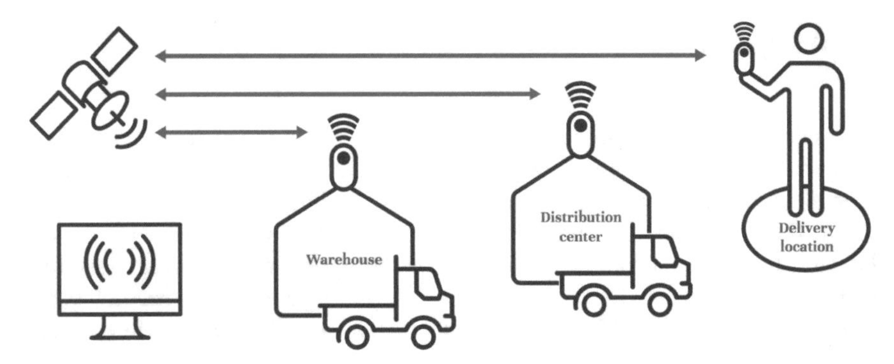

Fig. 2.4 Geolocation

Satellites are also used for location tracking, specifically those two dozen that form the Global Positioning System (GPS) sent to space by the US Department of Defense. Collecting data points at unique places and times can provide aid actors certainty that goods or vehicles have reached certain locations, even when relying on third party deliveries. Some actors already developed 'FedEx-style delivery systems' including low-tech options that use barcodes attached to packages with staff in different locations scan when the item passes them. This data, visualized on maps that are intuitive to understand, is valuable for real-time awareness of where to target which needs and has increased flexible decision-making and adjustments of aid efforts (Fig. 2.5).

Radio remains the most widely used technology reaching the largest number of people in remote areas around the world, especially in insecure environments. Broadcasts can be used to circulate important announcements but also to explain aid efforts and feedback mechanisms to crisis-affected communities. Radio programming itself can be used for active two-way engagement, involving or supporting communities in creating their own shows and stations. This can provide interesting forms of gathering feedback in and of itself, which some aid agencies are pioneering. Still, in humanitarian programming and monitoring especially, radio has not received mainstream

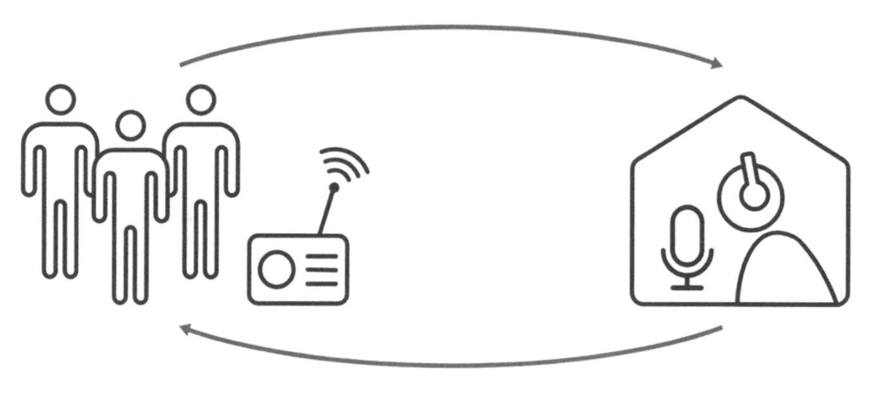

Fig. 2.5 Technology Type 4—Radio

attention. Radio is often seen as a one-way outreach tool, which is not intuitively ideal for M&E. However, a number of projects have shown that radio is easy to use and can complement feedback mechanisms and enable aid actors to seek new forms of input. It offers great potential for aid accountability and community resilience.

Wrapping up with this fourth reliable and likely underestimated tool, the wealth and variety of various technologies, with many news ones yet to come, is evident.

2.3 Digital Disasters

Implementation errors and inherent flaws of ICT

As so often, benefits and opportunities do not come without a dark side. Worse, in insecure environments the risks and challenges technologies introduce can be detrimental, even deathly. A small but growing literature started pointing to shortcomings and problems with advances in technology being brought to humanitarian and conflict contexts. Many of these challenges are surprisingly non-technical: Simply put, technology often fails when introduced too quickly in the wrong setting and for the wrong reasons (Sandvik et al. 2014; Jacobsen 2015). Transplanting what works in one part of the world or even humanitarian programming is no guarantee for success elsewhere. Where new tools are poorly coordinated or not discussed with other actors in the area or sector, proliferation and competition can undermine accountability efforts. Misjudgements of the effort, capacity and time required to maintain and run new systems can lead to mishaps, delays and inconsistencies. Similarly, where technical issues need to be resolved at headquarter level, delays and problems can occur. Finally, technical failures, especially if unanticipated, can severely hamper technology projects. Further yet, these critiques do not address more inherent challenges entangled with ICT and digitalization, a significant selection of which is described and categorized below.

Challenge 1: Mishaps and mistakes

Serious concerns came up in interviews with 'half-baked' implementation or unanticipated non-technical problems. In one case, aid recipients were given mobile vouchers directly to their phones. The aid organization, however, was not aware that telephone signals in the area were poor, requiring recipients to seek locations, for example, mountains, where their phones could connect to the network. Mountains, then, become perfect target spots for crime and loitering, when individuals would wander off alone in search of mobile connection. Similarly, where enumerators were asked to record GPS-stamps with their surveys, several reported having to wait minutes, sometimes hours, until they could connect to the satellite connection and thus move on to the next question. In another case, aid organizations sent tablets to their field staff but did not consider the number of plugs needed to charge each of them. 'I stayed up all night charging one iPad after the next,' a staff member said. Mobile hotlines, as another example, were sometimes not introduced well enough to fulfil intended purposes, so aid recipients would often call with questions or with praise rather than complaints. An even greater problem was where advertisements outlived hotlines, causing the community to call with nobody responding.

Challenge 2: Negative impact of new devices

Not all staff not always reacted well to new tools, and sometimes rejected change. Local enumerators hired for surveying often required more training than anticipated if they were handling digital devices for the first time, and needed detailed instructions including on the swipe motion to operate smartphones. For data collectors, restrictive digital survey forms and reporting platforms sometimes were a problem: certain answers did not fit. Small answer boxes on small smartphone screens also could cause difficulties with data entry and frustration. Some also critiqued new types of data power dynamics: where data is uploaded to a central database but servers are located at headquarter level and connectivity is poor, it is sometimes impossible or difficult for people in field locations to access the data they submit, either because of access barriers, or simply because the Internet is much too low to dial into the central system. An equally unwelcome inherent technology issue was introducing gender bias. Radio, for instance, is often known to cater more to male than female listeners, a trend that can be hard to break, so the humanitarian intervention could reinforce unequal access. Similarly, men are more likely to hold on or own the phones a household uses, making it difficult for women to use phones anonymously.

Challenge 3: Loss of trust and reputation

Beyond operational difficulties, aid staff observed that the sheer introduction of new devices could have negative effects. Many reported that local communities did not respond well: enumerators with smartphones instead of paper surveys were found to be less friendly, and hotlines were found to be less personal. In fact, respondents were frequently less willing to discuss sensitive issues over digital media, sometimes for the fear of interception or spying. In Syria, for instance, it was impossible to record GPS-stamps as armed groups would immediately suspect a connection to a foreign military and worry that their armed bases could be revealed. In Somalia, al-Shabaab banned aid actors with smartphones altogether and would threaten those who used them. The negative stigma of digital devices could also affect the quality of survey responses and information passed on via mobile phones. With regards to remote sensing, some governments banned aid agencies from using aerial imagery, fearing that the NGO could reveal data that could harm the government.

Challenge 4: Digital vulnerabilities and digital harm

Further, there are severe risks related to digital security that are far from addressed, let alone known, by both the humanitarian and other sectors. Even though aid practitioners often were aware that digital data can be copied and intercepted without them noticing and that online accounts can be compromised, mobile communications can be tracked and anonymized datasets can be reidentified, many organizations fail to prioritize digital security and even opt against using encryption to secure their data or devices. 'If militants knew how to intercept our communications,' one respondent said, 'then they would build better bombs.' And 'we offer encryption, but aid organisations never want it,' said two different service providers. Increasing digitization, though, means increasing dependency on tools, which both reduce redundancy. Any potential blackout or attack on digital databases or systems will likely become ever more costly. At the same time, it makes attacks more rewarding for attackers as they can get their hands on more information, which can be used against people or passed on to third parties. The potential cost of digital attack was far from meeting with adequate security precautions, typically for a simple reason: as long as no serious incidents are known, aid actors have other worries.

Challenge 5: Privacy risks and irresponsible digital data

Notably, where multiple datasets exist that include information about the same group or individual, for example, census data, food aid distribution charts, caller ID records and online social media accounts, these could allow digital attacker to reidentify a person. Comparing multiple datasets makes it possible to distil accurate revealing information about people. Increasing amounts of digital data, that are increasingly shared with digital communities around the world, for example, in order to 'crowd-source' the analysis of images, can unintentionally increase the risk of data abuse and cross-referencing. All in all, humanitarians' lack of interest in 'cyber' threats is alarming. If aid actors digitize more of their data and communications, they urgently need to increase their digital security efforts. Though some actors are developing promising protective tools, aid organizations overall might be well advised to listen to a quote from IT-security circles: 'There are two types of organizations: those who have been hacked, and those who will be.'

Challenge 6: Increasing inequality and power imbalances

Even those technology implementations that seem seamless and successful can entail problematic power dynamics that disadvantage aid recipients or lead to long-term negative consequences (Jacobsen 2015; Toyama 2015). Technologies affect their surroundings, which can be detrimental in contexts of discrepancy and donor–recipient relationships. In academia, technology has been critiqued as 'deepening the processes of creating inequalities' as those in power who introduce new tools risk undermine the engagement of those in the periphery (Santos 2000).

Challenge 7: Dependence on non-humanitarian actors and sectors

Technology-enabled aid attracts and actively depends on new actors to the field, including computer experts and for-profit businesses. Some of these may not adhere to, or even contradict, the humanitarian principles of neutrality, independence and impartiality (Raymond and Card 2015b). Increasing dependence on the services and expertise of these new actors can compromise the values and objectives of aid efforts (Duffield 2014).

Challenge 8: Double-standards and hypocrisy compromise humanitarian principles

Some controversy has risen where aid efforts relied on tools that were not considered ethical or adequate in other contexts (Hosein and Nyst 2013; Jacobsen 2015: 10). In Europe, for example, public backlash led policymakers to halt the integration of advanced biometrics in citizen registration. In contrast, this is widespread and praised for refugee registration (Hosein and Nyst 2013: 8; Jacobsen 2015). Aid actors often do not have regulatory safeguards for technologies (Duffield 2014: 3). Unless crisis-affected communities understand and object to the risks, subjecting them to digital technologies can introduce new dependencies and inequalities.

2.4 Mitigation Measures

Risk awareness aids responsibility

Oftentimes, the most successful tactics to minimize technology harm are not technical at all, but behavioural such as self-imposed limitations on what information to collect and transmit digitally (Steets et al. 2016; Antin et al. 2014; Internews 2015). When the name or age or even gender of a person is not critically necessary for household nutrition surveys, it may be better not to record it. Similarly, it is good to aggregate and generalize information at the first stage of data collection to prevent reidentification that could lead to harm or problems for individuals (de Montjoye et al. 2014). Security audits can help identify who has access to aid data and how easily staff accounts and emails could be compromised (Shostack 2014; Internews 2015). It is also good to introduce privacy-conscious technology and 'free and open source software (FOSS).' These tools run on code that is published and can be reviewed by anyone, such that security can be verified. Such tools can also be continued after individual projects may close down. Finally, aid actors should always plan for nondigital alternatives when connectivity or electricity fail, or ICTs cannot be used for other reasons. Sometimes, the best solution may be to opt against technology tools altogether.

Notably, mitigation happens at different levels. It can take involve impromptu measures in the field, or standards and good practice can be developed at headquarter level. Some outspoken actors are already seeking to establish shared norms across the field, including the 'digital development principles,' which were developed with about a dozen international and UN organizations and endorsed by many more (https://digitalprinciples.org). In addition, several initiatives emerged around 'responsible data' efforts, crafting principles and practical recommendations on how to assure data security and adequate access (https://responsibledata.io/). And in the UAV community, a code of conduct was authored and agreed (werobotics.org/codeofconduct/). Contributing to this work, Table 2.2 presents nine core ideas on implementing technology tools based on a 'backwards' analysis from risk assessment first to mitigation measures suggestions. This list is not necessarily comprehensive, and specific recommendations would depend on specific technologies and contexts. Yet, this first offering can help start conversations and be further refined with practitioners.

Back to low-tech—and sometimes no-tech

Broadly speaking, there are four types of circumstances when it may be better not to use technologies for humanitarian purposes, as the inherent risk would outweigh or overshadow possible benefits (Table 2.3).

Specifically, the following recommendations apply to the different technology types:

When using mobile phone-based systems, do not…

- *…use phone-based systems to collect sensitive data that could put beneficiaries at risk*. Information related to gender-based violence, the location of persecuted people or financial and health information that could cause stigmatization cannot be reliably secured when transferred through phone networks.
- *…use it for short-term projects or without continuity*. The set-up and familiarization costs only pay off if phone-based systems are used over a long time or for several projects.
- *…create a new mechanism where other, similar mechanisms already exist or are planned*. With too many systems in place, aid recipients can be confused and are less likely to use any.
- *…use it if you do not have the capacity to process feedback and follow-up*. Beneficiaries will expect follow-up when using phone systems. Lack of response can harm trust and reputation.
- *…simply replace other monitoring or feedback approaches*. Phone systems are not sufficient in themselves as phone ownership is biased and network coverage uneven.

Table 2.2 Risk mitigation for humanitarian ICT uses: why and what

1. Study the context before choosing tools	The information and technology ecosystems determine whether a new tool or approach can have a positive impact in the long term. Hotlines, for example, only work where locals have and use phones. Satellite images only make sense when skies are clear and smartphones should only be used in culturally appropriate ways *Be very clear about the information you need to gather or spread. Assess closely what type of information and knowledge travels via which channels in your context. Understand who influences and spreads information and can impact it*
2. Involve all users actively	Any tool or technology is only as effective as those meant to use it understand - and use - it. This involves all: programme staff, supervisors, data collectors, local communities, those processing the data and those making decisions based on it *Work with representatives of the different user groups when inventing, designing and testing the tools. Focus groups or interviews and, as much as possible, collaboration can help assure that ICT is usable and appropriate in all ways, including handling, pricing, language, etc.*
3. Establish informed consent practices	Achieving informed consent is especially difficult when aid actors themselves do not know all the risks involved with technologies and digitization. Because there is currently little best practice, aid actors need to handle recipients' data carefully *Agree on mechanisms and standards by which to explain the risks involved with handling survey responses or phone requests digitally. Do this well before disaster hits*
4. Provide back-up channels and alternatives	Technology-based efforts need to be prepared for the worst cases including energy outages, network disruption, theft of devices, software jams and other complications *Have analogue alternatives in place to turn to when the new tool does not work. Also, assure that every online function has an offline option. And do carry extra batteries*
5. Use security-conscious, free and open source software	With technology, intuition and observation are not enough to know who can gain access to information, as calls and electronically submitted data can be intercepted, often unnoticeably. The responsibility to safeguard personal data and keep it away from third parties lies with the aid organizations, so the choice of technology matters *Use only those tools that independent security experts can review. Such 'free and open source software' (or FOSS) options exist for moth relevant ICT tools*
6. Minimize and self-limit data	Even with the best tools, ensuring digital security is extremely difficult, also for experts. It is safe to assume that every data point that has been digitized can be copied or stolen. Even seemingly harmless datasets can reveal people when combined with other data *Collect only on a 'need to know' basis. Be clear which information gaps you are trying to fill and identify which data points you need to collect. Similarly, define access levels clearly. Who needs to see individual records and where do aggregate numbers suffice?*

(continued)

Table 2.2 (continued)

7. Invest in building acceptance	Local authorities can ban technology tools and restrict aid access. Armed groups could even target them. Local communities may lose trust if they do not trust the technologies *Plan trainings and meetings with local staff, authorities as well as community members. Explain what you are using and let them see and perhaps test the tool themselves*
8. Pool funds and risk	Technology implementations typically take more time and high investments. Running out of budget half-way and halting projects can create inconsistency and further risk trust *Collaborate with other aid actors in the area of relevant private companies. Share the investment in tools and seek agreeable mechanisms for sharing them and the data*
9. Apply humanitarian principles to technology	The humanitarian principles were developed long before the risks affiliated with digital communications were known. Their maintenance in online context must be explored *Safeguard the principles when working with new tools and partners. What does neutrality mean in the context of algorithms that sieve through information? How can you maintain independence when working together with private sector companies? Are biases towards those willing and able to use phones, for example, overshadowing universality?*

Table 2.3 Four scenarios when technology should NOT be used, but low-tech or analogue methodologies are recommended

Data is so sensitive that it could put people at risk	Acceptance is low and could hamper efforts: bans, suspicion, stigma, etc.	Infrastructure makes it impossible or costly: network connectivity, low spread of phones	Capacity constraints mean it cannot be guaranteed in the long term

When implementing digital data entry, do not…

- *…use digital devices, Internet or phone networks if they are in any way banned or compromised in the targeted area.* It is not worth the benefit if carrying the tool could lead to expulsion or mistrust and false accusations.
- *…use smartphones or tablets if they are very uncommon or clash with cultural norms and standards.* If the devices can create distrust or suspicion among local communities, they should not be used. Their stigma could hamper data entry.
- *…use where there is no phone or internet connection and/or electricity at all.* Lack of connectivity which would make it challenging to make the most from the tool.

When working with satellite imagery or geo-location tracking, do not…

- *…use satellite or UAV imagery and GPS-tracking if no guidelines on their use are in place and the potential risks to local communities are high or unclear.* Records of the location of vulnerable, persecuted or starving populations or of food delivery trucks and critical infrastructure can seriously endanger people and individuals if in the hands of their enemies.

- *…work with UAVs or other remote sensing technologies if local stakeholders seriously object to their use.* Remote sensing technologies can be associated with spying. Using them against the will of local authorities or communities can erode trust and put operations and staff at risk.
- *…invest in technologies where weather or context conditions are prohibitive and projects and their effects cannot be seen from the sky.* Satellites cannot see through clouds and in some instances regulation might inhibit use. Similarly, where projects can be seen, but there is no clear idea what visually observable impact would be expected, remote sensing may not be a worthwhile investment.
- *…commit to remote sensing when costs for images and analysis are disproportionate to the overall budget.* As of yet, evidence for overall gains or improved decision-making thanks to satellite use remains limited. Until then, their use should only be considered where costs are proportionate to overall project budgets and the expected gains.

When broadcasting information via *radio, do not…*

- *…broadcast information on radio when it reveals the location of vulnerable populations.*
- *…set up new radio programmes when you cannot guarantee long-term commitment to cater to the need of your listeners.* It is not worthwhile to build and audience if the show soon ends.
- *…use radio to support monitoring efforts when you cannot combine it with other tools* and approaches to receive responses to issues raised and questions asked in the shows.
- *…invest in independent radio stations or shows when you cannot engage local communities.* Their help in designing and running radio programmes is critical to assure local interest and relevance.

2.5 Looking Ahead

Scrutiny on using ICT in conflict settings can shed light and insight relevant to the broader field of working with technologies for humanitarian, development and human right purposes. However, the field of practice and literature on technologies used for humanitarian purposes and to communicate with local communities needs to be expanded to examine and address risks and mitigation measures. The particular focus on ICT for monitoring aid in access-constrained insecure settings in this paper provided a lens to view some of the practical and ethical concerns with using technologies to handle data of vulnerable populations. Other technologies relevant to humanitarian action include hardware (such as 3D printers and biometric ATMs), medical research (such as genetically modified food and new vaccines), social media (particularly platforms, messaging apps, and so-called ‘big data’) and a range of applications for coordinating, planning and implementing aid delivery. Many of the concerns around digital and physical risks outlined here similarly apply to these technologies. In the future, findings from the research underlying this paper as well as alerts and suggestions from other sources should be investigated further both through analysis and cross-comparisons with other fields and through practical sessions, workshops and discussions. These can contribute to the growing field of a responsible and sustainable use of technology for the benefit of conflict- and disaster-affected communities. They can inform aid efforts that safeguard the humanitarian principles and in their work 'do no harm' and do no digital harm either.

References

Altay, N., & Labonte, M. (2014). Challenges in humanitarian information management and exchange: Evidence from Haiti. *Disasters, 38*(Suppl. 1), S1–S23. https://doi.org/10.1111/disa.12052.

Antin, K., Byrne, R., Geber, T., van Geffen, S., Hoffmann, J., Jayaram, M., et al. (2014). *Shooting our hard drive into space and other ways to practice responsible development data*. Book Sprints.

Belden, C., Surya, P., Goyal, A., Pruuden, P., Etulain, T., Bothwell, C., et al. (2013). *ICT for data collection and monitoring and evaluation: opportunities and guidance on mobile applications for forest and agricultural sectors*.

Demombynes, G., Gubbins, P., & Romeo, A. (2013). *Challenges and opportunities of mobile phone-based data collection: Evidence from South Sudan*. Policy Research Working Paper No. 6321. The World Bank.

de Montjoye, Y. -A., Kendall, J., & Kerry, C. F. (2014). Enabling humanitarian use of mobile phone data. *The Journal of the American College of Dentists*. Retrieved from http://www.brookings.edu/research/papers/2014/11/12-enabling-humanitarian-use-mobile-phone-data.

Dette, R., Steets, J., & Sagmeister, E. (2016). *Technologies for monitoring in insecure environments*. Available at: http://www.gppi.net/fileadmin/user_upload/media/pub/2016/SAVE__2016__Toolkit_on_Technologies_for_Monitoring_in_Insecure_Environments.pdf.

DFID. (2013). *DFID's anti-corruption strategy for Somalia*. Available at: https://www.gov.uk/government/uploads/system/uploads/attachment_data/file/213926/anti-corruption-strategy-so.pdf. Last accessed January 12, 2016.

Duffield, M. (2013). *Disaster-resilience in the network age access-denial and the rise of cyber-humanitarianism*. DIIS Working Paper 2013: 23.

Duffield, M. (2014). *Cyber-intelligence in humanitarian emergencies: A critical exploration*. Notes. http://www.iss.nl/fileadmin/ASSETS/iss/Research_and_projects/GGSJ/Crisis_Speaker_photos/Speakers_Papers/Duffield.docx. Accessed November 20, 2015.

Gilman, D., & Baker, L. (2014). *Humanitarianism in the age of cyber-warfare: Towards the principles and secure use of information in humanitarian emergencies*. OCHA Policy and Studies Series 011. UN OCHA.

GSMA (2012). *Disaster response: Guidelines for establishing effective collaboration between mobile network operators and government agencies*. Resource document. http://www.gsma.com/mobilefordevelopment/wp-content/uploads/2013/01/Guidelines-for-Establishing-Effective-Collaboration.pdf. Accessed November 20, 2015.

Hallow, D., Mitchell, J., Gladwell, C., & Aggiss, R. (2012). *Mobile technology in emergencies: Efficient cash transfer mechanisms and effective two-way communication with disaster-affected communities using mobile phone technology*. Retrieved from http://www.alnap.org/resource/10619.

Hosein, G., & Nyst, C. (2013). *Aiding surveillance: An exploration of how development and humanitarian aid initiatives are enabling surveillance in developing countries*. London: Privacy International.

Hussain, H. (2013). *Dialing down risks: mobile privacy and information security in global development projects*. Washington, D.C.: New America Foundation.
Jacobsen, K. L. (2015). *The politics of humanitarian technology: Good intentions, unintended consequences and insecurity*. Abingdon: Routledge.
Kalas, P., & Spurk, C. (2011). *Deepening participation and improving aid effectiveness through media and ICTs*.
Korenblum, J. (2012). *Mobile phones and crisis zones: How text messaging can help streamline humanitarian aid delivery*. Available at: http://odihpn.org/magazine/mobile-phones-and-crisis-zones-how-text-messaging-can-help-streamline-humanitarian-aid-delivery/.
Madianou, M., Longboan, L., & Ong, J. C. (2015). Finding a voice through humanitarian technologies? Communication technologies and participation in disaster recovery. *International Journal of Communication, 9*(S.l.), 19. ISSN 1932-8036. Accessed November 20, 2015.
Maxwell, D., et al. (2014). *Lessons learned from the Somalia famine and the greater horn of Africa crisis 2011–2012: Desk review of literature*. Feinstein International Center. Available at: fic.tufts.edu/assets/Desk-Review-Somalia-GHA-Crisis-2011-2012.pdf. Last accessed January 12, 2016.
Raftree, L., & Bamberger, M. (2014). *Emerging opportunities: Monitoring and evaluation in a tech-enabled world*. New York: Rockefeller Foundation.
Raymond, N. A, & Card, B. L. (2015a). What is "humanitarian communication"? Towards standard definitions and protections for the humanitarian use of ICTs. Signal Program on Human Security and Technology, Harvard Humanitarian Initiative.
Raymond, N. A., & Card, B. L. (2015b). Applying humanitarian principles to current uses of information communication technologies: Gaps in doctrine and challenges to practice. Signal Program on Human Security and Technology, Harvard Humanitarian Initiative.
Raymond, N. A., Davies, B. I., Card, B. L., Achkar, Z. Al, & Baker, I. L. (2013). While we watched: Assessing the impact of the satellite sentinel project. *Science & Technology*, 1–13.
Robinson, A., & Obrecht, A. (2016). *Using mobile voice technology to improve the collection of food security data: WFP's mobile vulnerability analysis and mapping*. Available at: http://www.alnap.org/resource/21596.
Sandvik, K. B., Wilson, C., Karlsrud, J. (2014). A humanitarian technology policy agenda for 2016. Blogpost. Humanitarian Academy at Harvard (ATHA). http://www.atha.se/content/humanitarian-technology-policy-agenda-2016. Accessed November 20, 2015.
Shostack, A. (2014). *Threat modeling: designing for security*. Indianapolis, IN: Wiley.
Steets, J., Sagmeister, E., & Ruppert, L. (2016). *Eyes and ears on the ground: Monitoring in insecure environments*. Available at: http://www.gppi.net/fileadmin/user_upload/media/pub/2016/SAVE__2016__Monitoring_aid_in_insecure_environments.pdf.
Thakkar, M., Floretta, J., Dhar, D., Wilmink, N., Sen, S., Keleher, N., et al. (2013). *Mobile-based technology for monitoring & evaluation*. Retrieved from http://www.theclearinitiative.org/mobile-basedtechnology.html.
Toyama, K. (2015). *Geek Heresy: Rescuing social change from the cult of technology*. New York, NY: PublicAffairs.
UAViators. (2015). *Humanitarian UAV code of conduct & guidelines*. Resource document. Humanitarian UAV Network. https://docs.google.com/document/d/1Uez75_qmIVMxY35OzqMd_HPzSf-Ey43lJ_mye-kEEpQ/edit?pli=1. Accessed November 20, 2015.
Van de Walle, B., & Comes, T. (2015). On the nature of information management in complex and natural disasters. *Procedia Engineering, 107,* 403–411. https://doi.org/10.1016/j.proeng.2015.06.098.
Vazquez, L. R., & Wall, I. (Eds). (2014). *Communications technology and humanitarian delivery*. European Interagency Security Forum (EISF).
Woods, A. K. (2014). *Do civil society's data practices call for new ethical guidelines?*. Stanford: Ethics of Data in Civil Society.

Chapter 3
Unmanned Aerial Vehicles for Environmental Monitoring, Ecological Conservation, and Disaster Management

Raj Madhavan, Tatiana Silva, Flavia Farina, Rodrigo Wiebbelling, Leonardo Renner and Edson Prestes

3.1 Introduction

Unmanned Aerial Vehicles (UAVs) or Drones have received unprecedented attention in the last few years. From Amazon's ambitious plan of its fleet of prime air package deliveries right to our doorsteps to United States Air Force Reapers firing hellfire missiles at targets in Afghanistan and Iraq to paparazzi's chasing celebrities to get that prized picture of a Hollywood star's baby, they have conjured up images only limited by our vivid imaginations. Whatever mode of control they are flown under (they can be either autonomous or remotely piloted by ground operators), for a destructive or a constructive mission, UAVs evoke connotations in the minds of the public that borders on the absurd to highly valid concerns. Based on a recent study, it is expected that the market value will be a whopping USD 89 billion within the

T. Silva · F. Farina · R. Wiebbelling · L. Renner
Basin Modeling Laboratory, Institute of Geosciences, Federal University of Rio Grande do Sul, Porto Alegre, Brazil
e-mail: tatianasilva@ufrgs.br

F. Farina
e-mail: flavia.farina@ufrgs.br

R. Wiebbelling
e-mail: rfwiebbelling@gmail.com

L. Renner
e-mail: cardoso.renner@ufrgs.br

R. Madhavan (✉)
Humanitarian Robotics and Automation Technologies, Clarksburg, USA
e-mail: rajmadhavan.t4h@gmail.com

E. Prestes
Phi Robotics, Institute of Informatics, Federal University of Rio Grande do Sul, Porto Alegre, Brazil
e-mail: prestes@inf.ufrgs.br

S. Hostettler et al. (eds.), *Technologies for Development*,
https://doi.org/10.1007/978-3-319-91068-0_3

next 10 years (Drones: Market Overview 2014) Our take on UAVs is that they can be an invaluable tool and an asset in the sky providing images of unsurpassed quality in civilian applications ranging from disaster relief, environmental monitoring, and surveillance to combating animal poaching efforts and situational assessment and awareness.

Extreme events became obvious in Brazil in 2011 when the mountain region of Rio de Janeiro faced extreme rain in the order of 270 mm overnight. High slopes, clay soils, irregular settlements, and enormous amounts of rain lead to thousands of deaths and render dozens of thousands homeless (the official count is 35,000 but the actual number is much higher than that). This was the final outcome of the major hydrometeorological disaster in the Brazilian records. Rio Grande do Sul in Southern Brazil consists of several high-risk areas. Moreover, according to the Brazilian Civil Defense, almost 600 extreme events led municipalities to declare a state of emergency in Rio Grande do Sul between 2003 and 2010. Almost a million people were affected, either by becoming homeless or missing, getting injured, sick, dying, or having to relocate or in extreme cases, resulting in fatalities. It is not a matter of time before we are faced with floods and landslides. It is already happening. It is a matter of time until they happen regularly in such a scale that we start referring to them as disasters. Recent events lead us to the obvious conclusion: we were not prepared for that anywhere in Brazil. So the question is: how can we prepare, react, and adapt to this new environmental reality?

Among the diversity of natural environments around the world, certain ecosystems can be considered as vital areas, given their environmental functions in food production, hydrological control, sheltering, genetic stocking, etc. Hence, it is mandatory for our own survival to keep them protected to ensure the preservation of life support systems. However, this is not an easy task since significant manpower is needed to patrol thousands of hectares of land and water to control illegal fishing and hunting. In order to track global changes and their impact on the Earth system, we need specific tools and instruments, e.g., land use and cover mappings which are essential to model flood and landslide risks. A common way to perform these mappings is to use the information gathered by orbital remote sensing and/or cameras carried by manned aircraft. While the latter involves high costs and complicated logistics, the former cannot be used to support an emergency plan when a hydrological disaster occurs since its use imposes constraints like fixed revisit frequency; large elapsed time between an observation and the image supplied; and the presence of clouds.

An interesting and efficient solution is to use UAVs for obtaining a bird's-eye view of the landscape and also for capturing detailed and high-resolution images of the ground. UAVs can acquire high-resolution multispectral images, which in turn can produce 3D models of hazard zones without putting human lives at risk. Meteorologists, hydrologists, and spatial modelers can use such data to provide better detail and reliable results at a local scale. In this context, UAVs have the potential to assist in the acquisition of spatial data and also to evaluate the behavior, distribution, and size of the populations under study.

Our work with UAVs is centered on creating tools to support the preservation of human lives and the ecosystems that support them. In Taim Ecological Station located in Southern Brazil, Federal University of Rio Grande do Sul (UFRGS) and other stakeholders have acted collaboratively to assess geographic information to help the elaboration of an environmental plan to solve specific community demands and also to monitor the impact and dynamic aspects of the ecosystem, such as the occurrence of fire, invasive species, and environmental infractions.

We have developed Geographic Information Systems (GIS)-based regional models of environmental risk and how they overlap infrastructure and human settlements in the Rio Grande do Sul coastal zone (Silva et al. 2011; Silva and Tagliani 2012). Our current work focuses on enhancing the regional models using precision and super high-resolution images taken by UAVs especially for those areas already identified as high risk. The greater the precision of these models, the more lives can be saved because these models can form the basis for public policies in constraining human settlements on high-risk areas. Moreover, meteorological sensors carried by UAVs can be employed to collect atmospheric information, providing better precision for the meteorological models at a local level. When these data are integrated in hydraulic/hydrological models, scene-generation becomes possible, thus allowing us to predict which regions are vulnerable to floods or landslides depending on different levels of rainfall. This information can then aid rescue teams when a certain level of rainfall is expected once it is incorporated in collaborative platforms such as participatory WebGIS or similar ones.

This article details the development of an integrated framework that merges remote sensing capabilities (including but not limited to UAVs) using GIS models in a collaborative fashion with social capital. To the authors' knowledge, our efforts represent the first instance of the merging of multi-resolution remote sensing data, WebGIS, policy, and disaster management practices to arrive at an analytically rich, practical, and effective national disaster response structure for Brazil and possibly extendible to other developing economies. This article describes our work with UAVs for applications in environmental risk assessment, ecological monitoring, landslides/floods detection, and prediction, as well as regulatory aspects in Southern Brazil.

3.2 Environmental Vulnerability and Susceptibility Studies

Even if our eyes see the world in front of us as a colored 3D structure, it is much more than that. It is multidimensional. A single spot on the Earth's surface will have a certain level of vegetation density or it may be covered by urban settlement or buildings. And underneath it, you will find a certain type of soil, a water table at a certain depth, and, if you dig deeper, a certain type of rock. This spot will present an elevation value related to the sea level. And it may not be at the same plane related to as the sea level either but it can be present at a certain slope. Depending on the direction this slope-face (called "aspect"), this spot will receive a certain amount of sunlight and be more or less exposed to the wind forces. And all of these characteristics vary along the 3D space and through time. Besides that, the way water will flow when it rains depend on every one of these characteristics or dimensions.

Part of it will infiltrate into the soil, part of it will run off, and part of it will evaporate. In a high slope spot covered by concrete, the water will run off. In a spot covered by a forest on a sandy soil at a low slope, water is going to infiltrate easily. So if a forest is replaced by agriculture, or even worse, by concrete, less water will infiltrate and, consequently, more water will run off. If water flows on the surface, it carries soluble substances, grains, small pieces of rock, big pieces of rock, cars, houses, and people. However, it is not enough to think in terms of a single spot, but we need to consider a region instead, and each one of these dimensions (vegetation, land use, soil, geology, slope, etc.) as superimposed layers.

GIS is the conventional platform to cross as many layers of information as needed so we can detect and model the spatial relationships between parameters. Once we formulate the (map) algebra, we can then generate digital maps depicting runoff and flow accumulation which accumulation, which, in turn, indicate landslide and flood risk. As noted earlier, spatial criteria change through time. We can use the same geology and soil maps as long as we live because they will not change for hundreds or thousands of years. But the same cannot be said about land use and cover. Forests turn into pasture. Natural fields turn into rice fields. Human activities are changing land cover ever faster. So fast that when a land use and cover map is done it is already outdated!

Remote sensing enables us to be quick in assessing this process. Sensors carried by satellites give us quantitative measures of the way a material interacts with the electromagnetic energy so we can differentiate targets based on what we call *spectral response pattern*. If a satellite carries a radar sensor, or yields stereo pairs, we can also obtain topographic data. This type of information comes as images, so we can mathematically manipulate them and find ways to automatically detect land use and cover changes. However, when we need very detailed spatial information, sometimes urgently, satellites present several constraints: low spatial resolution of free/low-cost images, high costs related to high spatial resolution images, a fixed revisit frequency, and requirements to schedule an observation, the time lapse between an observation and the image supply, or the simple presence of clouds.

What to do when we need to be very sure whether an extreme meteorological event will be hazardous or not for specific infrastructures or populations? High-resolution land use and cover maps, as well as 3D models, are required for hydrological, hydraulic, and meteorological models at a very local scale. In the imminence/occurrence of an extreme event, rapid assessment of high-risk/affected areas is needed. It would be way too cumbersome and labor-intensive to regularly map land cover and use and build 3D models with a high resolution (in the order of 2–5 m) for a region as large as 60,000 km^2. This is the actual area of the Littoral Basin in Rio Grande do Sul where our efforts are focused (see Fig. 3.1). Besides, when it comes to map land use and cover, which is driven by human activities, we are pretty much always out of time. GIS models based on satellite data are great. They give us a broad view of vast regions so we can detect high-risk areas as well as where flood and landslide risks are increasing over time.

Our current work, thus, focuses on enhancing the regional models using precision and high-resolution images acquired using UAVs (Fig. 3.2), especially for those

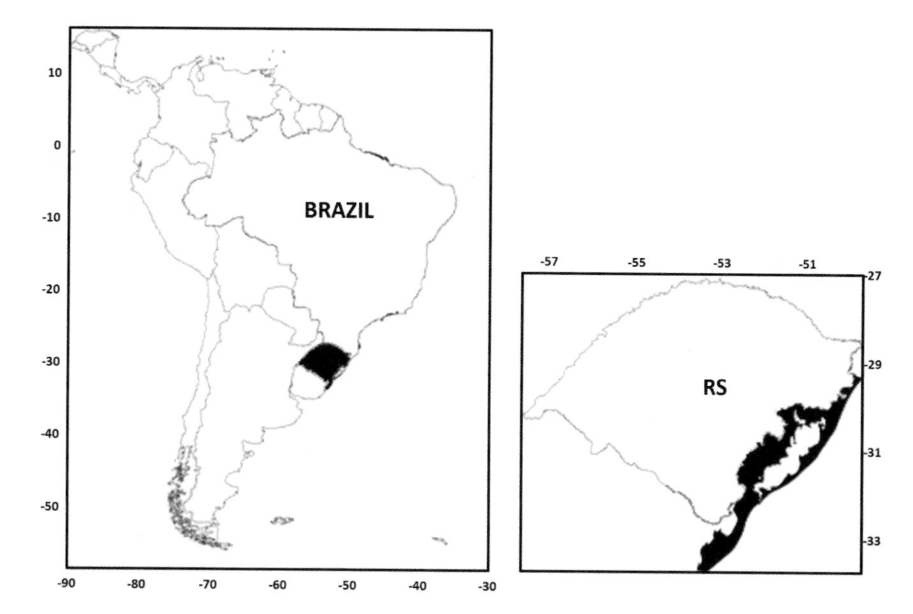

Fig. 3.1 Southern Brazil coastal plain

areas already identified as high risk. Robust, fixed wing UAVs are preferable in this phase as the goal is to acquire quality images as fast as possible. Satellite images with equivalent quality are costly and not so easy/fast to obtain for aforementioned reasons.

We deploy the ECHAR 20A UAVs developed by XMobots (XMobots), a Brazilian UAV manufacturer. It comprises very desirable characteristics: it flies up to 3 km altitude under wind speeds up to 45 km/h and low rain, has 45 min. autonomy, cruising speed of 75 km/h, stall speed of 37 km/h, visible RGB resolution 2.4–19.2 cm. The challenge lies in the now relies in the image processing since ECHAR 20A yields 24 bits images with about 35,106 pixels per image. A 3D model based on images acquired using the UAVs provides the accuracy necessary to build detailed hydrological/hydraulic models.

The greater the precision of the input data, the greater the precision of the models, and consequently more lives can be saved because these models can be the basis for public policies in constraining human settlements on high-risk areas and to set off evacuation alerts. Moreover, meteorological sensors in the ECHAR 20A can be employed to collect atmospheric information thus providing better precision for the meteorological models at a local level. When these data are integrated into hydraulic/hydrological models, scenario generation becomes possible, as well as to predict which regions floods or landslides, depending on the levels of rainfall. This information also helps to anticipate the action of rescue teams when a certain level of rainfall is expected.

Fig. 3.2 A false color-composite image obtained by ECHAR 20A. Land use and cover mapping can be done with precision based on these images

The cataclysmic landslide in Rio de Janeiro provided some valuable lessons: Communication and field assessment are crucial. UAVs can replicate signals and enable communication when telecommunication towers collapse. UAVs can also be instrumental in planning alternative routes when roads and bridges are destroyed, and to (remotely) find people in need.

3.3 Ecological Conservation: Mapping and Target Detection

Land use and cover changes lead to soil impermeabilization. But we need land for agriculture, urban settlements, etc. Human disrupted ecosystems cannot replace these vital areas. They simply cannot play the same roles. Conservation units normally consist of vital areas. That means, ecosystems that play a vital role for the environmental functioning. Some of them are totally untouched; others have human uses more heavily regulated. It is not an easy task to monitor a conservation unit. It commonly means thousands of hectares of land and water to monitor and patrol. UAVs can provide the necessary information to better observe nature and to detect and prevent illegal activities, such as hunting and fishing.

Taim Ecological Station in Southern Brazil is an integral conservation unit. This unit spans a very wide wetland, internationally recognized by the RAMSAR

Fig. 3.3 Left to right: Lab-made fixed wing, Phantom 4, and DJI S-900

Convention (RAMSAR Convention). We have flown trial flights in this region but already have started to set mapping procedures to assess geographic information not available so far to support the decision making in two different situations:

(1) to help elaborating an environmental plan, which depends on mapping many spatial parameters with a high resolution, and
(2) to monitor impacts and dynamic aspects of the ecosystem. For the latter case, we have identified the need to build our own UAVs.

Figure 3.3 shows some of the UAVs often used in monitoring the Taim Station. Commercially available UAV platforms are usually not flexible enough for covering wide range of applications as those needed in managing and maintaining a conservation unit (e.g., land cover change detection, fauna and flora identification and monitoring, fire detection and monitoring, illegal fishing, and hunting detection). Besides, "lab-made" UAVs can reduce costs by more than ten times especially if a swarm of them need to be developed for distributed coverage of the vast land area. Since conservation units in Brazil can only rely on limited funding, it is imperative to bring forth affordable options of UAVs. Lab-made UAVs are being developed and is presented in last section.

3.4 Regulatory Constraints in Brazil

Regulation is a crucial aspect of UAV usage in Brazil. UAVs weighting more than 150 kg are under the same regulations of airplanes. UAVs between and 150 kg require a Brazilian Aeronautics Register and a formally habilitated pilot in performing any flight. UAVs with less than 25 kg need a register in the SISANT (Sistema de Aeronaves Não-Tripuladas/Unmanned Aerial Vehicles System), and also a habilitated pilot in performing flights above 400 feet. Flights above 400 feet also need to be authorized by ANAC (Agência Nacional de Aviação Civil/National Agency for Civil Aviation). Only UAVs under 250 g do not require anything to fly. Above that, UAVs must keep a minimum distance of 30 m from people, the pilot must be older than 18 years, and acquire a third-party damage insurance. In any situation, DECEA (Departamento de Controle do Espaço Aéreo/Aerial Space Control Department) rules must be followed. People close to the flight area must be aware and authorize the flight. Pilots of UAVs with 25 kg or more need a CMA (Certificado Médico Aeronáutico/Aeronautical Medical Certificate) emitted by ANAC or CMA. The use of UAVs for loads or goods transportation is forbidden. Only electronic devices such as cameras or computers are allowed during flight. The portage of the following documents is mandatory: flight manual, risk assessment document, and insurance. All these requirements and restrictions make it complicated to start applying UAVs in any field in Brazil, especially with bigger sensory payloads, which in general are more suitable for mapping purposes.

3.5 Ongoing and Future Work

GIS-modeling capabilities for UAV images, both for ecological conservation and disaster management, are currently being enhanced. For the Taim Ecological Station specifically, 3D models have been built, and some other applications are under study. The suitability of different UAVs for many management demands in Taim has been tested. These include capybara automatic counting (for population dynamics studies), fishing and hunting detection, nest detection, and bird identification. As mentioned before, low-cost UAV is the most desirable one for these cases, similar to the lab-made ones that have been built (Fig. 3.3). The final goal in this sense is to define and create the most suitable "UAV family" to cover all types of technology gaps related to monitoring.

Social networks have emerged as an obvious way to exchange information during disasters. SMART (Simulation, Modeling, Analysis, Research, Teaching) Infrastructure Facility, part of the University of Wollongong in Australia, has used Twitter to trigger disaster alerts. Facebook now sends automatic messages enabling the user in a disaster area to post updates on his/her status. The next step in our work consists of integrating UAV data into such existing frameworks, as well as providing relevant spatial information accessible through WebGIS platforms.

References

Drones: Market Overview. (2014). http://www.robolutioncapital.com.
RAMSAR Convention. The Convention on Wetlands. http://www.ramsar.org/.
Silva, T. S., De Freitas, D. M., Tagliani, P. R. A, Farina, F. C., & Ayup-Zouain, R. N. (2011). Land use change impact on coastal vulnerability: subsidies for risk management and coastal adaptation. In *Proceedings of International Symposium on GIS and Computer Mapping for Coastal Management* (pp. 54–59).
Silva, T. S., & Tagliani, P. R. A. (2012). Environmental planning in the medium littoral of the Rio Grande do Sul coastal plain Southern Brazil: Elements for coastal management. *Ocean and Coastal Management, 59,* 20–30.
XMobots. http://www.cpetecnologia.com.br/topografia/vant/547/echar-20a/xmobots.

Chapter 4
Higher Education Spaces and Protracted Displacement: How Learner-Centered Pedagogies and Human-Centered Design Can Unleash Refugee Innovation

Barbara Moser-Mercer, Erin Hayba and Joshua Goldsmith

4.1 Introduction

Today, the world is seeing the largest number of refugees and displaced persons in history, with a total of 65.5 million forcibly displaced and over 22 million refugees in 2016 (UNHCR 2016). The average conflict lasts 10 years, and families are displaced for an average of 20 years. Against a backdrop of competing humanitarian needs and priorities, the education response is often limited, leaving entire generations uneducated, developmentally disadvantaged, and unprepared to contribute to their society's recovery (UNHCR 2011).

Traditionally, Education in Emergencies responses have focused on primary education; higher education opportunities have often been perceived as a luxury. Current statistics on refugee access to education confirm this ongoing trend: 50% of refugee children access primary education, 22% secondary education, and only 1% higher education. Children and youth are particularly vulnerable to losing their right to education, a basic human right that is enshrined in the 1989 *Convention on the Rights of the Child* and the 1951 *Refugee Convention*, and is essential to the exercise of many other human rights. In 2015, the United Nations adopted the Sustainable Development Goals, thereby broadening the education mandate to include lifelong learning.

While refugee youth have extremely limited options in conflict and crisis zones (Sheehy 2015), rapid advances in technology, online learning and open educational resources (OERs) have laid the foundations for making higher education opportunities accessible for refugee youth.

B. Moser-Mercer (✉) · E. Hayba · J. Goldsmith
University of Geneva, InZone, Geneva, Switzerland
e-mail: Barbara.Moser@unige.ch

E. Hayba
e-mail: erinhayba@gmail.com

J. Goldsmith
e-mail: Goldsmith.joshua@gmail.com

S. Hostettler et al. (eds.), *Technologies for Development*,
https://doi.org/10.1007/978-3-319-91068-0_4

Education fosters innovation and entrepreneurial skills that are important for employability, economic activity, and job creation—elements that are critical for stability during times of reconstruction and for longer term sustainable development. If refugees and internally displaced persons receive a quality education while in exile, they are more likely to develop the necessary skills to make use of the existing economic, social, and political systems in their host communities as well as upon returning home.

4.2 Background

Over the years, InZone—a Higher Education in Emergencies (HEiE) actor whose mission is to design, develop and scientifically validate HEiE models that respect humanitarian principles—has identified key HEiE challenges: access to education, quality of educational programs, relevance of programs, and management of programs in the field and remotely.

During an initial mapping exercise of higher education providers active in the refugee education space, quantitative data was collected using an online questionnaire and disaggregated. This data presented a snapshot of existing programs in terms of several key features, such as size, graduation rates, funding, scholarship availability, accreditation, types of courses, degrees and fields of study, learning spaces, and use of technology. In the second phase of the research, in-depth interviews were conducted with educators responsible for running a number of such higher education programs. Interviews were analyzed using computer-assisted qualitative data analysis (CAQDAS) programs and a series of inductive codes and constructs which emerged holistically from the data (cf. Bryman 2004; Corbin and Strauss 2007; Merriam 2009).

Cross-cutting themes that emerged during this mapping exercise included cultural dimensions, intellectual and learning cultures, the development of twenty-first-century skills, peer-to-peer and collaborative learning, pedagogical models that integrate ICT and connectivity, responsible partnerships, information sharing and communication, leveraging refugees' skills and ingenuity, and preventing brain and knowledge drain.

When designing programs for fragile contexts, most HEiE providers have aimed to facilitate equal access to programs. Nevertheless, a series of physical, intellectual, financial and legal factors have hampered this access, including insufficient infrastructure and connectivity, the length and cost of travel to the learning center, insecurity, a lack of ICT skills, language barriers, religious and cultural factors, course costs, and a lack of needed legal documentation, particularly education credentials, to access higher education. Providers have adopted a wide range of innovative solutions to surmount these challenges.

Though HEiE programs attempt to survey the contexts in which they will be operating before launching their programs, it generally takes several editions of a course to develop approaches which dovetail with the reality on the ground. HEiE providers often offer programs which they believe would be of use to refugee learners, such as teacher training courses, yet these models, which are generally incubated in the Global North, must be adapted so that they are accessible for and meaningful to refugee learners and provide options for learning pathways.

One of the main questions to emerge from these reflections is just how refugee learners can learn in such a context. How can learners communicate with institutions of higher learning and with organizations willing to support refugee access to higher education? How can learners share their needs across languages and cultures, including learning cultures, particularly if the gaps in understanding are too deep and too wide? How do providers foster cultural and linguistic ownership and ensure that courses are culturally and linguistically relevant?

The main challenge for higher education providers entails escaping the lure of tried-and-true higher education models. How can we avoid the attractiveness of the concept of humanitarian assistance as a way to "provide"? How can we rise to the challenge and recognize that the greatest successes shall be found in a participatory approach that engages and empowers beneficiaries to co-design solutions and make learning a two-way process? Indeed, the true beneficiaries of all Higher Education in Emergencies initiatives must be refugee learners and their communities, since learning is both an individual and a social activity, and the knowledge and skills refugee learners acquire should also help advance their communities.

4.2.1 Disruptive Higher Education in Fragile Contexts: Digital Ecosystems and Open Educational Resources (OERs)

Even in traditional Higher Education (HE) models, education is not synonymous with taking a course or pursuing a course of study, whether in a lecture hall, a seminar room or on a learning portal. Education includes knowledge and skill acquisition, instruction, debate, application of acquired expertise, critical inquiry, cultural expression, and transmission to other members of the community and society.

Refugee youth enrolled in Higher Education programs also seek learning environments that challenge them to learn and grow beyond the textbook or classroom. While the liberty to think and challenge ideas and opinions is often the norm in universities, such space is limited or nonexistent in refugee contexts. In times of crisis, insecurity, fear, and destruction, youth are greatly affected and have few opportunities and even less space for free and open thinking. Furthermore, refugee youth are often too old to qualify for most educational programs and too young to have much experience or stability to help them through such difficult times (Evans and Forte 2013). Living in dangerous conflict and crisis areas poses numerous challenges for refugee youth, including the threats of armed groups and tight legal restrictions on employment and freedom of movement (Burde 2014).

Higher Education thrives in larger spaces that offer diverse inputs and encourage free expression, allowing learners to combine free navigation of learning resources with more structured tutoring and mentoring approaches. Unorthodox learning pathways unleash creativity; interrupted learning pathways, however, are hurdles. Protracted displacement is the biggest such hurdle, a major stumbling block to satisfying intellectual curiosity and leading a dignified life. This is due to the feeling of helplessness and insecurity that can compound feelings of despair and hopelessness over years of displacement with limited or no educational or economic opportunities.

Open Educational Resources (OERs), and more specifically Massive Open Online Courses (MOOCs), have disrupted Higher Education in the Global North by challenging traditional assumptions and practices about teaching and learning, most notably by putting the learner at the center of the design and using technology to enable the learning process. As such, MOOCs have acted as a catalyst for change and have prompted many educators to rethink what education should look like today and how we can best design it.

Through our ongoing research into the challenges of refugee access to higher education (Moser-Mercer 2016), we have learned that despite extraordinary growth in ICTs, access to information in fragile contexts remains constrained by connectivity, language, and cultural barriers. Furthermore, information per se does not constitute education, and this is particularly true for learners living in fragility. Higher Education in Emergencies cannot simply replicate traditional Higher Education models. Indeed, HEiE must unlock the innovation potential of its "users" in an uncertain environment by operating flexibly while showing tolerance for ambiguity, adaptability, and equifinality, i.e., researching multiple pathways to reach its objectives (Ries 2011; Blank 2013; Bloom and Betts 2014).

This perspective motivated InZone to undertake the MOOC Multi-Center Study (Moser-Mercer 2016). This project aimed to engage in a finer grain analysis of the multiple factors that contribute to the development of optimal learning environments in fragile contexts. These factors must be responsive to refugee learners' needs, respect humanitarian principles, and create opportunities for charting new learning pathways in situations of conflict, crisis, and protracted emergencies.

The study built on an earlier case study (Moser-Mercer 2014) in an effort to strengthen the evidence base for HEiE by including a variety of fragile learning contexts. These differed with respect to the setting—camp or urban refugee settings; language—multilingual with English as vehicular host country language or predominantly monolingual with Arabic as vehicular host country language; and gender diversity. Even after enlarging the learner pool, expanding into different urban- and camp-based refugee settings, ensuring diversity in the participant pool and incorporating refugees as participant researchers, the earlier case study results were largely confirmed.

The study used an off-the-shelf MOOC offered consecutively as a session-based and on-demand course in English. It integrated as variables (a) on-site and virtual tutoring and mentoring support in select settings, enlisted the partnership of NGOs and international organizations with an education mandate and local presence, and (b) two geographic regions—the Horn of Africa and the Middle East—and two vehicular languages—English and Arabic. The research aim was to increase our understanding of the variables which mediate successful learning outcomes for all refugee learners. Especially when it came to the variable of tutor support, the qualitative analysis generated particularly useful results designed to inform the adaptation of OERs in general, and MOOCs in particular, to fragile contexts. In fact, through quantitative analysis of on-site tutoring support the latter emerged as the single most decisive variable to boost learner success, particularly for female refugee learners in refugee camp settings. The study carefully documented the differential use of social media and learning technologies and concluded that fragile learning contexts make use of different modes of modern communication and that choices are usually dictated by connectivity, accessibility, and financial affordances.

This confirmed the relevance of the Technology-Access-Matrix, designed by InZone to support all of its online and blended higher education programming (InZone 2015). This matrix contends that higher education courses for learners in fragile contexts should be designed to integrate a carefully planned progression regarding the use of learning technologies and communication media in an effort to provide pedagogically sound incidental learning opportunities. This contrasts with the more traditional approach to HEiE, whereby ICT courses are offered prior to launching content-rich and engaging higher education programs, thus artificially separating the acquisition of ICT skills from the actual skill of learning. This approach compartmentalizes skills that together are essential for supporting learning, making it more difficult for learners to understand their respective relevance and hampering their required cognitive integration. Skill and knowledge acquisition are co-determined by learners' capacity, learning opportunities, motivation, and engagement. In challenging learning contexts, integrative learning pedagogies are vital to building successful learning strategies.

Support courses, such as language enhancement or ICT courses, often do not elicit the kind of engagement from learners in general, and refugee learners in particular need to maintain motivation in the face of contextual adversity. This spans a wide range of challenges, including insufficient, irregular access to learning resources that are not explicitly made available by course providers in a durable format, but which learners are expected to access; and language and cultural barriers that represent cognitive challenges that are not easily resolved in a virtual learning environment.

The MOOC Multi-Center Study offers rich qualitative and stringent quantitative data to inform course providers about learner support requirements. These need to be integrated into course planning if large numbers of refugee learners are to succeed and remain motivated to continue their learning journey. While this study did not specifically address the question of credentialing OERs and allowing learners engaged in non-formal higher education to obtain regular academic credit for their work, the different requirements for session-based and on-demand versions of OERs and the way learners navigated these provide important information as to how such credentialing may have to be approached in the future. Of particular note are the issues related to hard deadlines, peer-assessed learning activities, learning ethics—including the interpretation of the concept of plagiarism across different cultures—and free-loading. In the refugee context, learning goes global. Nevertheless, humanitarian principles, humanitarian accountability, and guidelines for education in emergencies have recently been developed by the Connected Learning in Crisis Consortium (http://connectedlearning4refugees.com) to ensure that all voices are heard and all cultures, including intellectual cultures, are respected.

Learning from and designing with higher education learners thus emerges as the optimal approach to HEiE, as it puts the refugee learner in the center of the design process, builds on bottom-up processes of evidence-generation, leverages humanitarian partnerships in an effort to meet humanitarian principles and guidelines, and thus promotes participatory innovation. Innovation is already part of the humanitarian system. It is driven by a demand for new models and rapid technological change (Bloom and Betts 2014). User-linked and user-led or "indigenous innovation" is by definition inclusive and thus has the potential to foster ingenuity among youth living in fragility contexts.

Since OERs deliver knowledge, encourage experimentation with information, and invite adaptation to local contexts and languages, they are both "adoptable" and "ownable." Such innovation is often termed frugal, grassroots, or BoP (bottom-of-pyramid); HEiE actors consistently point out the challenges of designing for, with and by users, yet unanimously agree that such an approach opens up many innovative possibilities for not only finding solutions to problems but also ensuring that these solutions are compatible with the local context and thus have a higher chance of successful diffusion (Smith et al. 2013). As Bloom and Betts (2014, p. 28) aptly state, "…researchers and entrepreneurs from outside the traditional humanitarian agencies benefit from collaborating with end-users and agencies to define problem statements and designs." Although bottom-up innovation in fragile contexts is subject to humanitarian, legal, economic, and social constraints, there is considerable space for engagement and "innovation spaces," which can be physical or virtual spaces for learning, interaction, and sharing ideas and resources. These spaces are enabled by a digital ecosystem that supports HE models offering quality education opportunities and livelihoods and encourages safe, non-physical ways of resolving conflict, thereby addressing a major humanitarian challenge.

Learning and designing from and with our learners is indeed the bottom-up approach that has continually informed InZone's research: When learners reflect on their challenges on a regular basis, they not only improve their meta-cognitive skills but are ultimately empowered to be the agents of the change they wish to bring about. Research and pedagogy are thus intimately intertwined; every study carried out in fragile contexts should also incorporate a pedagogical agenda. This will discourage traditional research approaches—where refugees are merely the subjects of research and subsequent solutions and models are imposed on them—and encourage a participatory approach that engages refugees as active research contributors and designers of their own spaces.

4.2.2 Designing HE Spaces in Settings of Protracted Displacement: The Case of the Kakuma InZone HE Space

InZone launched higher education courses in Kakuma Refugee Camp (located on the border between Kenya and South Sudan) in 2012, and in Dadaab Refugee Camp (located on the border between Kenya and Somalia) in 2013. Within the larger methodological framework of a case study, and for the purposes of this article, the authors present an incremental empirical approach to designing HE spaces in fragile contexts while observing humanitarian principles. The Kakuma InZone HE Space represents a case study that pools qualitative and quantitative data collected across different refugee camp settings and explores the potential of collaborative pedagogies to address protracted conflict.

While written within a non-refugee context, Radcliffe's (2009) statement on designing education spaces can be readily transposed to fragile contexts. According to Radcliffe, a learning space should motivate learners and promote learning as an activity; it should support individual, collaborative, and formal practice and provide a personalized and inclusive environment. Furthermore, it should be flexible, adapting to changing needs and contextual challenges and to the chosen pedagogical approach. The more challenging the context, the more important the design and the greater the degree of flexibility that is required to let users adapt different parts of the space to support different dimensions of their learning.

Radcliffe's pedagogy-technology-space framework rests on a number of important assumptions about designing optimal learning environments. First, pedagogy is enabled by space, which in turn encourages pedagogical innovation. Second, space embeds technology, which in turn expands the space and opens it up to entire communities of learners, both locally and globally. Third, pedagogy benefits from technology as an enabler, while technological innovation is enhanced by pedagogical considerations. Higher education spaces for refugee learners should thus be flexible, to accommodate current and evolving pedagogies; future-proofed, to enable reallocation and reconfiguration of space; bold, to look beyond tried and tested technologies and pedagogies; creative, to energize and inspire learners and tutors; supportive, to develop the potential of all learners; and enterprising, so that each space can support different purposes.

In order to implement InZone's collaborative, learner-centered pedagogy—which is embedded in its framework of skill acquisition and expertise—InZone is piloting the design of a Higher Education Space for Refugees in Kakuma Refugee camp (InZone HE Space Kakuma). The design incorporates twenty-first-century approaches to technology-supported learning in an extremely challenging learning context, a refugee camp. The design is bold in that it looks beyond traditional approaches to higher education by integrating both formal and non-formal learning opportunities; access to credit-bearing and diploma-granting higher education programs is blended with non-formal higher education programming. This includes MOOCs, which are offered in both English and French, as well as on-site workshops that leverage the potential of the arts—dance, movement, applied drama, storytelling, and creative writing—to allow learners to be creative and to unleash their creative potential to co-design their learning environment and amplify the knowledge acquired through higher education opportunities through sharing it with the entire refugee community. It is also future-proofed, by allowing for reallocation of space—the Kakuma InZone Student Café hosts student discussion groups, community performances, and learning groups—and can expand to other parts of the camp, since the Café is geographically separate from the InZone Learning Hub. The design is flexible, adapting to complementary pedagogical approaches and enabling both individual and collaborative learning; it is enterprising, in that each space can support different learning activities; and most of all, it is supportive, continuously encouraging learners to creatively use their wider learning space and thereby encouraging all HE refugee learners to develop their potential to the fullest.

The concept of an HE space in fragile contexts thus covers and enables the entire range of learning possibilities. It begins with structured spaces for traditional face-to-face teaching, online and blended learning in the InZone Learning Hub, where learners can either work individually or collaboratively with or without tutor support and thus with more or less pedagogical structure. It encompasses debates in the InZone Café, operated by Ethiopian refugee learners who ensure that traditional coffee ceremonies are passed on to learners in exile. It extends all the way to less structured learning spaces, such as the French library, and ultimately includes the home environment. Learning is therefore not confined to a specific location or to a specific environment. Rather, learning happens in a range of contexts, some of which are more structured than others. Incidental and formal learning are both equally encouraged. The HE space transitions easily across contexts, as do refugee learners, who often hold multiple jobs in order to eke out a living in the camp. Most importantly, though, the Kakuma InZone Higher Education Space is always open to redesign and configurable to the needs of different higher education programming. It is open for collaboration with other higher education providers in the camp. Consequently, it contributes to establishing a community of higher education learners, who are equipping themselves with twenty-first-century skills and thereby contributing to making their communities conflict-resilient as they actively search for peaceful alternatives to the wars they have left behind. Most importantly, the Kakuma InZone Higher Education Space is refugee managed through a team of refugee learners and InZone alumni, rather than through an implementing partner, as is traditional in humanitarian and development projects. Skills and competencies acquired through higher education thus support employability and secure livelihoods, thereby empowering refugees to design and redesign the learning space in careful synergy with their community.

4.2.3 Learner-Centered Pedagogies and Human-Centered Design in Support of SDG 4 and EFA

In 2015, the Sustainable Development Goals (SDGs) were adopted by the United Nations to expand upon the Millennium Development Goals which concluded that same year. SDG 4 focuses on education and reads, "Ensure inclusive and quality education for all and promote lifelong learning." This SDG redefines and shifts the approach to education set out in the previous MDG on education: learning is now considered a lifelong process, and is not limited to primary education. According to the most recent UNESCO Global Monitoring Report on Education for All (EFA) in 2014, great progress was made in recent years towards reaching the targets laid out in the MDGs, particularly for universal primary education. Progress towards targets in secondary education access was also made. Yet in areas of conflict, the number of out-of-school children has increased (UNESCO 2015).

Humanitarian education programs have typically focused on children, particularly those enrolled in primary education, as educational programming in humanitarian contexts is decidedly reliant and linked to international policies (Zeus 2011). Yet, with the new shift towards lifelong learning embedded in SDG 4, humanitarian education programs will need to be more holistic, broadening their approach to youth and higher education. International humanitarian and development policies play an instrumental role in determining education programming on the ground in conflict and crisis settings, setting the stage for institutions of higher learning to engage as humanitarian actors.

The global targets laid out in the SDGs refer to a broad mandate of deliverables by the year 2030. Among these, specific mention is given to tertiary education; increasing the number of youth and adults with skills for employment and entrepreneurship; educating children and youth in vulnerable situations; and promoting peace and appreciation of diversity. These targets, while ambitious, will require innovative solutions and changes to current educational programming in emergency contexts. More importantly, the learners in these communities should be active contributors to the design and implementation of such programs. Human-centered design will thus emerge as a critical aspect of this process, which can be further promoted through learner-centered pedagogies that have been an integral part of the InZone approach and which prepare refugee learners for lifelong learning journeys.

In designing and producing OERs with other HEiE actors in the North and the South, InZone has taken its lead from refugee learners themselves, engaging them in the design process and supporting bottom-up innovation. Evidence from the implementation of OER-driven courses informs the creation of new resources, particularly with respect to how to design "the last mile" to optimally support learning outcomes. Higher Education programming for fragile contexts should be field-proofed and co-designed by learners before it aspires to scale. To meet the objectives of SDG 4 and EFA, we must listen to beneficiaries. By engaging them in the design of both content and delivery, we will foster communication across languages and cultures in the field, ensure that learning materials are contextualized and access to learning is guaranteed, and pilot promising solutions together before going to scale. In fragile contexts, beneficiary input to all phases of Higher Education provision is thus essential and needs to be integrated into all SDG-4 programming and funding schemes.

4.3 Conclusions

This paper has analyzed the contribution of Open Educational Resources to building twenty-first-century skills; noted the value of tutoring and mentoring models for optimal learning outcomes, learner retention and the provision of language and subject matter support; and explored the technologies, including learning technologies, that best mediate higher level learning in fragile contexts. Variables such as sustainability, operability, equal access, cultural and linguistic ownership, livelihoods, and context relevance were used to analyze available evidence in an effort to inform optimal design and scalability of such learning spaces, as well as their potential use in migrant refugee contexts.

We conclude that human-centered design is central to both pedagogical and technological development in fragile contexts. In light of the fact that protracted displacement settings can differ considerably from one another, there is no substitute for adopting a bottom-up approach and developing learning spaces that allow for both intellectual and artistic expression. Traditional HE programs are usually geared to specific degrees, still offer little flexibility, and do not cater adequately yet to alternative learning pathways. Online education platforms offer a wide range of learning options and a variety of pedagogical approaches. Such innovative tools and programs increase access to learning and offer great potential for higher education; however, such initiatives may face resistance from traditional university programs, particularly due to challenges regarding accreditation, recognition of degrees, and credit transfers (Hullinger 2015).

OERs have the potential to inspire non-traditional learners and encourage an explore-and-develop approach rather than a listen-and-reproduce mentality. They are thus a gateway that allows refugees to unleash their innovation potential. We emphasize the importance of refugee ownership and empowerment as vectors for ensuring the sustainability of higher level learning in HE spaces in fragile contexts, analyze the potential of ubiquitous learning supported in broadly conceived and flexible learning spaces for fostering creativity and innovation, and encourage the inclusion of innovative and learner-centered pedagogical approaches that support such learning in the larger framework of Education for All and Sustainable Development Goal 4.

References

Blank, S. (2013). Why the lean start-up changes everything. *Harvard Business Review, 91*, 3–9.

Bloom, L. & Betts, A. (2014). *Humanitarian innovation: The state of the art*. Occasional Paper Series UN-OCHA, November 2014/009. Available from http://www.unocha.org and http://www.reliefweb.int.

Bryman, A. (2004). *Social research methods*. New York: Oxford University Press.

Burde, D. (2014). *Schools for conflict or for peace in Afghanistan*. New York: Columbia University Press.

Corbin, J., & Strauss, A. (2007). *Basics of qualitative research: Techniques and procedures for developing grounded theory* (3rd ed.). Thousand Oaks, CA: Sage.

Evans, R., & Lo Forte, C. (2013). *A global review: UNHCR's engagement with displaced youth*. Geneva: Switzerland.

Hullinger, J. (2015). This is the future of college. Last accessed November 12, 2017 from http://www.fastcompany.com/3046299/the-new-rules-of-work/this-is-the-future-of-college.

InZone. (2015). *Technology-access-matrix*. Unpublished internal InZone report. Geneva: University of Geneva/InZone.

Merriam, S. B. (2009). *Qualitative research: A guide to design and implementation*. San Francisco: Jossey-Bass.

Moser-Mercer, B. (2014). MOOCs in fragile contexts. In U. Kress & C. D. Kloos (Eds.), *Proceedings of the European MOOC Stakeholder Summit 2014 (EPFL Lausanne)* (pp. 114–121). ISBN 978-84-8294-689-4. Available from www.openeducationeuropa.eu.

Moser-Mercer, B. (2016). *MOOC multi-center-study 2015*. Unpublished Internal InZone report. Geneva: University of Geneva/InZone.

Radcliffe, D. (2009). A pedagogy-space-technology (PST) framework for designing and evaluating learning spaces. In D. Radcliffe, H. Wilson, D. Powell, & B. Tibbetts (Eds.), *Learning spaces in higher education: Positive outcomes by design* (pp. 9–16). Brisbane: The University of Queensland.

Ries, E. (2011). *The lean startup: How today's entrepreneurs use continuous innovation to create radically successful businesses*. New York: Randomhouse.

Sheehy, I. (2015). *Tertiary education in conflict and crisis: The new landscape. Lecture*. Last accessed November 11, 2017 from https://mediaserver.unige.ch/fichiers/view/89797.

Smith, A., Fressoli, M., & Thomas, H. (2013). Grassroots innovation movements: challenges and contributions. *Journal of Cleaner Production, 63,* 114–124. https://doi.org/10.1016/j.jclepro.2012.12.025.

UNESCO. (2015). *EFA global monitoring report 2015: Education for all 2000–2015: Achievements and challenges*. Paris: UNESCO.

UNHCR. (2011). *Refugee education: A global review*. Geneva: UNHCR.

UNHCR. (2016). *UNHCR statistics portal*. Geneva, Switzerland. Last accessed November 13, 2017 from http://www.unhcr.org/figures-at-a-glance.html.

Zeus, B. (2011). Exploring barriers to higher education in protracted refugee situations: The case of Burmese refugees in Thailand. *Journal of Refugee Studies, 24*(2), 256–276. https://doi.org/10.1093/jrs/fer011.

Chapter 5
Healthsites.io: The Global Healthsites Mapping Project

René Saameli, Dikolela Kalubi, Mark Herringer, Tim Sutton and Eric de Roodenbeke

5.1 Introduction

The world is facing an increasing number of complex natural or man-made humanitarian crises. In order to respond to these growing challenges, humanitarian actors are deploying more and more innovative technologies and approaches to support relief aid (Haselkorn and Walton 2009). Geo-information are an example where humanitarian practices have dramatically evolved in the recent years with the emergence of the phenomena called by Goodchild (2007) Volunteered Geographic Information (VGI), or more generally by Burns (2014) "digital humanitarianism". Indeed, the democratization of access to Global Positioning System (GPS), satellite imagery and web mapping platforms, such as OpenStreetMap (OSM), and more recently mobile phone data collection tools have enabled large numbers of remote and on the ground individuals to produce geographical information to support humanitarian action. This technological evolution has offered greater opportunities of use for humanitarian actors. It has enabled the collection and sharing of large amount of data in short period of time at a fraction of the costs of the traditional data collection and

R. Saameli (✉) · D. Kalubi
International Committee of the Red-Cross, Lausanne, Switzerland
e-mail: rsaameli@icrc.org

D. Kalubi
e-mail: dikolela.kalubi@epfl.ch; dkalubi@icrc.org

M. Herringer · T. Sutton
Healthsites.io, London, UK
e-mail: mark@healthsites.io

T. Sutton
e-mail: tim@kartoza.com

E. de Roodenbeke
International Hospital Federation, Bernex, Switzerland
e-mail: ederoodenbeke@ihf-fih.org

S. Hostettler et al. (eds.), *Technologies for Development*,
https://doi.org/10.1007/978-3-319-91068-0_5

map-making methods (Haworth and Bruce 2015). However, the full scope of opportunities offered by VGI is still underused by traditional humanitarian actors. In order to understand this situation, Richards and Veenendaal (2014) have analyzed comprehensively the gap between the United Nations World Food Programme Crisis Mapping Operations and Crowdsourcing Technology. They conclude that crowdsourcing captured a large amount of data, but not sufficiently the required ones for the agency operations and with not the needed quality. Haworth and Bruce (2015) also highlight this need of enhancing data quality assurance to enhance the relevance of VGI for disaster management. Data quality in VGI relies heavily on Linus's law, which implies the more observers, the more likely an error will be identified. As shown by Haklay et al. (2010) in their analysis of Open Street Map data quality, this law seems to work well for spatial accuracy. However, Mooney and Corcoran (2012) have also discovered serious quality issues with tags or annotate objects in OSM. Such attribute data is usually much more needed than accurate geographical coordinates to support humanitarian management.

Geo-information on health facilities in disaster areas are a good example of the challenges of the use of VGI for humanitarian action. The most comprehensive healthsite geodatabase based on VGI is probably OSM, but information on services offered are still largely incomplete and questionable in terms of reliability. Other health geodatabases with comprehensive set of helpful attributes for health workers exist, but these databases are not easily shared outside of the health organizations which have gathered them, or are only regional in their coverage. OSM and these restricted datasets complement each other in terms of geographical coverage and in terms of the information they contain, however they are almost never readily available in a consolidated, freely and accessible way. Data exchange between VGI communities and health organizations is usually unidirectional and punctual. Traditional humanitarian agencies tend to task digital communities only for specific tasks lasting a rather short period of time (Burns 2014).

In order to address this issue, the Global Healthsites Mapping Project has been launched in 2015 to create an online interactive map, Healthsites.io, of every health facility in the world and make the details of each location and services easily accessible. A team of freelance developers, researchers, the International Committee of the Red-Cross (ICRC), and the International Hospital Federation (IHF) have joined their competences and networks in order to provide a single point of reference for healthcare workers, aid agencies, contingency planners, government agencies, and citizens who need access to a highly curated global dataset of healthcare facilities. In order to meet this aim, the project team has to address three major challenges:

1. Integrate multiple unstructured datasets in one unique database
2. Enhancing the reliability of data
3. Foster sharing and updating of information

In Sect. 5.2, this paper will present the approaches that are currently developed to address these three challenges. In Sect. 5.3, the paper will analyze the potential impacts, risks, and the perspectives of this project.

5.2 Healthsite.io Approach

5.2.1 Datasets Integration

Due to its open data license, its large number of entities, its worldwide coverage, and its large community, OSM was chosen as the main dataset of the project. In addition, several databases from trusted partners were chosen. All this data gathered represented more than 150,000 health localities worldwide. In order to integrate all these datasets with different structures and values, a data model, based on the Entity–Attribute–Value model, and inspired by the OSM data model was chosen. The OSM data model enables users to store information about anything. A FullTextSearch index has been implemented to enable searching for textual data. However, this flexibility of storing information represented a risk potentially hampering the goal of having a curated database with easily accessible information. In order to address this issue, 15 core attributes (Table 5.1) that are relevant for both the public and health professionals were defined by the International Hospital Federation. For most of these core attributes, defined values were set to enable comparisons. Indeed, an essential question such as "What type of health facilities is it" can vary greatly according to the national or organizational classification system.

Each time a record is created in Healthsites.io, a globally unique identifier is created and assigned to that record. This is specific to the Healthsites.io database and is used to provide a canonical point of reference for that record.

The upstream id is used to create a back reference to the original source data (e.g., from a national healthsites dataset).

5.2.2 Validation Process

In order to enhance the reliability of health facilities data, several validation processes have been planned to be added to the project. There is an automatic verification ranging from a simple: "email address should look like an email address" to more complex which even rely on external services like: "check if the Locality address is similar to results returned by external geocoding services" or "check if the telephone number is correct by manually calling the number and verifying".

There will also be a validation process based on user reputation (Fig. 5.1) to assess the reliability of data through a Locality Validity Index (LVI). This reputation-based process consists of four steps. The first step takes place during data integration. Depending on source of the data, the LVI obtains a score ranging from 0 to 10; 0 being for data added by a new user, 10 for data from a community trusted user, and 5 if it is a batch of data coming from a Ministry of Health or healthcare organizations. The reliability of the user is based on the monitoring of his activity on the platform and the crosscheck of his activity by other users. Second, once the data is displayed on the map of Healthsite.io, any user can complete missing attributes data and modify

Table 5.1 HealthSite.io database core attributes

Healthsite attribute name	Value
uuid	Universally unique identifier
upstream_id	Unique identifier placed on the source location
date_modified	Date time stamp of data modification
name	Name of the facility local name and in English
geom	Lat/long
source	Name and or website of the organization supplying the data
physical_address	Address
contact_number	Number including international dialing code (+250 252 588888)
nature_of_facility	Clinic without beds, clinic with beds, first referral hospital, second referral hospital or General hospital, tertiary level including University hospital
scope-of-service	All type of services, specialized care, general acute care, rehabilitation care, old age/hospice care
operation	24/24 & 7/7; open only during business hours; other (specify)
inpatient-service	Number of full-time beds, number of part-time beds
ancillary-services	Operating theater, laboratory, imaging equipment, intensive care unit, emergency department
activities	Medicine and medical specialties, surgery and surgical specialties, maternal and women health, pediatric care, mental health, geriatric care, social care
staff	X full-time equivalent doctors and Y full-time equivalent nurse, level of competency of the health workers
ownership	Public, private not for profit, private commercial
raw_data_archive_URL	A link to the raw data file (.csv, etc.)

or validate existing attributes data through a tweet channel. Third, the more users confirm the validity of information, the higher the LVI becomes. Within this step, if an authoritative user such as a staff of health organization, verifies information, the LVI becomes even higher. Finally, if time goes without anyone validating the record, the LVI progressively decreases.

5.2.3 Updating

Over the last years, World Health Organization (WHO) have developed and supported several projects, systems and guidelines for national and regional authorities to map their health facilities. However, in many countries where humanitarian workers intervene, this data once collected is often not updated or remains not easily accessible. In

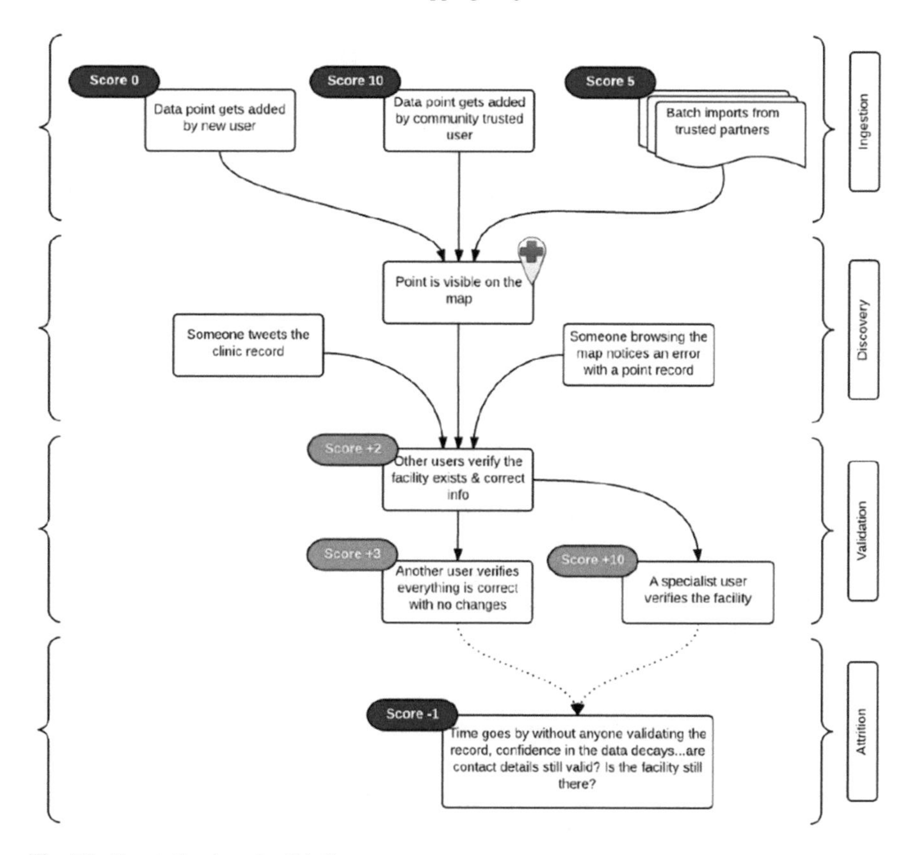

Fig. 5.1 Reputation based validation process

the absence of research on the cause this situation, one can only assume that authorities at regional and national level lack of resources to carry out the monitoring of healthcare services. In order to overcome this issue, the Healthsite.io project has been designed to have a bottom-up approach of the monitoring and data collection. Indeed, it is OSM community who provides the localization of health facilities and health workers in the field who provide related attribute information, such as the type of services available. These two sources of data are consolidated in Healthsite.io and data quality verified through the LVI. Once data quality is good enough, data are then sent back to OSM in order to be shared widely through the community of OSM users.

5.2.4 Opportunities, Risks, and Perspectives

This project aims to democratize access to health geodata. By enhancing the quality and accessibility of geodata, this project offers several opportunities to achieve social, health and humanitarian impacts in areas where data is scarce. A better knowledge of locations of health facilities and associated services can help to map and identify population underserved by healthcare services (Munoz and Källestål 2012) (Blanford et al 2012). Such information can contribute to develop and advocate better evidence-based health policy. It can also help healthcare organizations to better plan their activities, vaccination campaigns based on the correlation between vaccination rate and distance to primary healthcare center (Al Taiar et al. 2010). For relief workers, knowing where the health facilities and associated health service are located are crucial information to know to prepare contingency plans, but also to respond promptly and meaningfully during the emergency.

The project also bears some limitations and risks. Several studies (Haworth and Bruce 2015) have showed that questions about liability of VGI are often not clearly answered "Who is responsible if harm results from reliance on volunteered information: the initial contributor, the host or the organization responsible for the website or product relying on VGI". The question of moral and legal reliability is highly relevant in the case of public exposure and dissemination of potentially sensitive data such health facilities and workers location. In order to address this risk, a governance committee composed of ICRC, IHF and developer team has been set to monitor data quality and the risk of exposure of data in sensitive contexts.

The capacity of the project to meet its goal to build global, complete and high-quality database of world health facilities depends on the contributions of VGI communities and health experts. In comparison with other similar VGI initiative, this project has the advantage of having the endorsement of major actors such as Humanitarian OSM team, the ICRC, and IHF. This support enables a strong promotion among digital humanitarians and health experts.

Finally, the model of Healthsite.io of taking OSM data to enhance its attribute by health experts before exporting back completed and validated data to OSM is a truly innovative approach. It fosters bidirectional exchange between health experts and VGI communities. In addition, it also a new approach to quality issue of attribute data in OSM. If this model proves to be successful, it could be replicated in other essential domain, such water services, where accessibility to detailed and reliable data is much needed for humanitarian operations. This approach would be a new way to maintain updated comprehensive thematic database without having to spend resources usually required for such tasks

Research on how to measure quality of VGI, specifically attribute data, are recent and few. Healthsite.io offers an interesting concrete case to measure the reliability of data jointly validated by VGI community and authoritative experts. Such research could help to identify opportunities to enhance VGI validation process. In Healthsite.io, this process is mainly based on user reputation where also only few research on VGI applications have been done.

Furthermore, research on motivation of VGI and health experts contributing to such project could help to develop strategies to ensure long-term updating of the database.

References

Al-Taiar, A., Clark, A., Longenecker, J., & Whitty, C. (2010). Physical accessibility and utilization of health services in Yemen. *International Journal of Health Geographics, 9*, 38.

Blanford, J., Kumar, S., Luo, W., & MacEachren, M. (2012). It's a long, long walk: Accessibility to hospitals, maternity and integrated health centers in Niger. *International Journal of Health Geographics, 11*, 24.

Burns, R. (2014). Rethinking big data in digital humanitarianism: Practices, epistemologies, and social relations. *GeoJournal, 80*(4), 477–490.

Goodchild, M. (2007). Citizens as sensors: The world of volunteered geography. *GeoJournal, 69*(4), 211–221.

Haklay, M., Basiouka, S., Antoniou, V., et al. (2010). How many volunteers does it take to map an area well? The validity of Linus' law to volunteered geographic information. *Cartographic Journal, 47*(4), 315–322.

Haselkorn, M., & Walton, R. (2009). The role of information and communication in the context of humanitarian service. *IEEE Transactions on Professional Communication, 52*(4), 325–328.

Haworth, B., & Bruce, E. (2015). A review of volunteered geographic information for disaster management. *Geography Compass, 9*, 237–250.

Mooney, P., & Corcoran, P. (2012). The annotation process in OpenStreetMap. *Transactions in GIS, 16*(4), 561–579.

Muenoz, U., & Källestål, C. (2012). Geographical accessibility and spatial coverage modeling of the primary health care network in the Western Province of Rwanda. *International Journal of Health Geographics, 11*, 40.

Richards, S., & Veenendaal, B. (2014). *Understanding the gap between the United Nations World Food Programme crisis mapping operations and crowdsourcing technology*. Conference Article. Curtin University. http://espace.library.curtin.edu.au:80/R?func=dbin-jump-full&local_base=gen01-era02&object_id=235814.

Chapter 6
The Faceless Mobile Youth of Africa Drive Change

Darelle van Greunen and Alida Veldsman

6.1 Background

One of the major challenges facing most developing countries such as South Africa today is how to take full and smart advantage of quite spectacular and far-reaching advances in science and technology not only in promoting national economic development but particularly addressing the needs of poor and marginalised sections of society. Whilst technology access in the pre-digital era has always been a major barrier to economic and social development in developing countries, this is no longer an absolute barrier for at least three reasons: first, the proliferation of distributive and open technologies, and limits to the power of proprietary frameworks; second, the increasingly lower delivery cost of new technologies; and third, the emergence of multiple sources of technology innovation and development in civil society.

Despite this, it is not altogether clear that South Africa is sufficiently exploiting the multiplier effects of new technology opportunities across the traditional rural and urban, and black and white social barriers, especially insofar as real empowerment of the poor and marginalised are concerned. There is still, in some respects a growing, technological divide in our country, and the dividends of our technological age are rarely passed on equally to different social classes, ethnic groups and geographic communities in South Africa. The problem of a continued technological gap may be partly to do with the economics of 'access' by poor communities. However, there also seems to be major problems with the absence or weakness of requisite conditions—social, institutional and political—necessary for its effective social appropriation.

D. van Greunen (✉) · A. Veldsman
The Centre for Community Technologies, Nelson Mandela University,
Port Elizabeth, South Africa
e-mail: Darelle.vanGreunen@mandela.ac.za

A. Veldsman
e-mail: Alida.Veldsman@mandela.ac.za

S. Hostettler et al. (eds.), *Technologies for Development*,
https://doi.org/10.1007/978-3-319-91068-0_6

Technology has significantly changed the way in which young people interact with one another and the world around them. The youth use the Internet or a mobile phone to source information, engage, construct and maintain social networks. Technologies have dramatically transformed young people's relationships with one another, their families and communities. In South Africa, 75% of youth even in rural areas has access to a mobile phone (Eyebers and Giannakopoulos 2015). This not only yields opportunities for access to different types of information but also to create a faceless connected force of youth.

This paper sets out to explore the potential of using mobile phone technology to drive change amongst at-risk youth in a low-income community that is severely challenged by social issues.

6.2 Why the Northern Areas?

The Nelson Mandela Bay Metro is located on the southern coast of South Africa. The total population amounts to some 1.5 million of which some ±182,000 people live in the Northern Suburbs (Bloemendal, Bethelsdorp and Gelvandale) of Port Elizabeth (StatsSA 2011). Although the situation changes constantly as a result of socio-economic and other circumstances, it is estimated that there are approximately 40,000 households in informal areas with an additional 50,000 qualifying households in backyard shacks. Distorted development in Nelson Mandela Bay has manifested in a highly skewed distribution of income and wealth. The unemployment rate is over 35% of the total population (StatsSA 2011). At least 44% of households access at least one social grant with 20% of residents not having any or limited schooling.

Daily, the Northern Areas communities face some form of crisis, disaster or situation. The community members are currently downcast and stuck in a perpetual cycle of poverty, sickness, crime and hopelessness, and therefore do not progress but regress and are in need of bold moves to be taken towards poverty eradication, medical relief, unemployment eradication, crime eradication and development of the future leaders of not only this area but also the country as a whole. Gang culture has influenced life in the Northern Areas since the early 70s, with the birth of notorious gangs, such as the Mongrels, Mafias, 40 Thieves and Fire Boys. These gangs warred over turf and were greatly influenced by the code of the prison gangs. As more young men went through the prison system, and as gang groups split and splintered, more street gangs developed and gang culture became more ingrained.

The majority of young people who become involved with gangs are enticed by the hopes of financial gain, peer pressure or out of rebellion against parents and other forms of authority. Scores of young people, including pupils from the Northern Area schools, were murdered in violent gang-related attacks in recent years. Local press reports on gang violence and murders on a daily basis. According to the SAPS statistics between mid-2014 and January 2015, 59 gang-related murders were committed in the Northern Areas of Port Elizabeth; and 120 cases of attempted murder are being investigated. The police also claim that children as young as twelve are being lured into gangs and are groomed for the gang lifestyle.

Matavire (2007) describes Helenvale, a suburb in the Northern Areas of Port Elizabeth as a '*filthy, overcrowded and crime-ridden suburb… The Northern Areas township, which is predominantly coloured, and one of the oldest in Port Elizabeth, is characterized by extreme poverty and lack of service delivery. It has a population of more than 14,000, of which 75% have no income. The average income per household is estimated at R400 a month. At least 7% of the residents have no formal education. Overcrowding is a concern and, although most houses have toilets, they are shared between 10 and 20 people…The suburb had been planned during apartheid and the infrastructure is now falling apart- sewerage, housing and other facilities…*'.

6.3 Problem Description

The poverty experienced by these communities has left the members with low self-esteem. This is because they are unable to meet their basic needs. This population also lacks life skills that would enable them to meet their needs and perform their duties to the optimal capacity. This deprivation has led them to believe that they do not measure up to others who meet their family needs and lead quality lives. Educational underachievement occurs in every school in South Africa. It is caused by various factors and although it has negative consequences for the individual, the school, the community and the country as a whole, teachers and parents tend to simplify the problem by ascribing it to the 'laziness and stupidity' of the learners. This problem is emphasised in the Northern Areas of Port Elizabeth. An area challenged by severe social problems including poor health, substance abuse, domestic violence, gang fighting and teenage pregnancies. There is currently a big discrepancy in the number of children from the Northern Areas, starting school and those finishing grade twelve. The learners from the Northern Areas are academically underachieving. This is a problem because academic underachievement contributes to a vicious cycle of failure, behavioural problems, school dropout, unemployment, overcrowding, social problems and a general feeling of worthlessness. Children from an early age on a daily basis use drugs.

There is a need to understand what will empower the youth to drive change. In a society where people fear for their lives on a daily basis, there is a need to investigate other means of driving change. Currently, the social disconnect in the Northern Areas necessitates alternate forms of communication and support towards change.

6.4 Research Objectives

The objective of the research is to determine how Information Communication Technology (ICT) can be used to act as the driver of change without excluding human participation. What social structural factors relate to unique patterns in youth mobile phone usage in low-income areas faced with severe social challenges? We suggest that certain social dynamics inherent in the institutions of family, public places and peer relations are key factors. The youth use mobile phones because they enable new kinds of social contact, but also because different socio-economic environments are limited in access to adult forms of social organisation. This paper relies on ethnographic observations of members of a youth leadership academy as well as surveys and interviews conducted with different youths representing the Northern Areas community.

6.5 Methodology

Drawing on articles and reports by academics combined with ethnographic observations, this paper encompasses a range of disciplines including education, sociology, culture and information technology. Whilst the paper draws upon a variety of literature, the focus is on the South African context. An exploratory case study approach is adopted that applies interviews, surveys and direct observations as data collection mechanisms.

6.6 Mobile Youth Culture

Adolescents and young people have been identified as the first adopters of mobile technology with 72% of 15–24 year olds reported as having a cell phone in a national survey conducted by the Kaiser Family Foundation (2007) and the South African Broadcast Corporation in South Africa in 2007. Prepaid mobile subscriptions are cheap, do not require the user to have a bank account, physical or postal address and airtime can be bought as money becomes available. Together with services such as ‘Please Call Me’, which is a free notification service that a user can send to another person’s phone, and ‘WhatsApp’, which allows users to exchange messages without having to pay for SMS, made mobile phones the preferred communication tool among the youth. Since phones are usually handed-down to the junior members in a household, as the parents upgrade to newer and better devices, the problem of obtaining phones does not exist.

Where adults tend to have more conservative uses for mobile phones, young people find mobile phones quite liberating as it allows them access to just about anything they are interested in, anywhere, any time. To the youth today, mobile phones have become an extension of themselves. It is therefore not surprising that the youth uses their phones to interact socially via apps like Facebook Messenger, Pinterest, Instagram and WhatsApp to name but a few. Through Facebook, Twitter, MySpace and other social media networks, youths create virtual communities that not only give them a voice but also offer anonymity. Virtual communities are slowly replacing face-to-face communication. It would therefore be shortsighted not to utilise these new ways of communicating and interacting as a mechanism to drive change and build resilience among at-risk-youth.

The role of digital literacy and cyber safety is established, although policy and practice have been slow to respond to new ways of thinking about media literacy in a digital world. Traditionally, media literacy has been understood and taught in relation to mass media, addressing issues of media ownership, censorship and advertising. However, today's online and networked media environment requires a more complex digital literacy that is often not explicitly taught in school. This environment requires that young people develop new skills to participate and stay safe in the new digital media environment. Consequently, there are a number of components to technology literacy (Third and Richardson 2010). These include the following:

- *Technical literacy*—for example, the knowledge and skills required to use a computer, web browser or particular software programme or application;
- *Critical content literacy*—the ability to effectively use search engines and understand who or what organisations created or sponsor the information; where the information comes from and its credibility and/or nature;
- *Communicative and social networking literacy*—an understanding of the many different spaces of communication on the web; what is appropriate digital behaviour; level of privacy (and therefore level of safe self -disclosure for each); and how to deal with unwanted or inappropriate communication through them;
- *Creative content and visual literacy*—in addition to the skills to create and upload image and video content, this includes understanding how online visual content is edited and constructed, what kind of content is appropriate;
- *Mobile media literacy*—familiarity with the skills and forms of communication specific to mobile phones (e.g. text messaging/instant messaging); mobile web literacy, and an understanding of mobile phone etiquettes.

6.7 Social Media

Although people have been using the Internet to connect with others since the early 1980s, it is only in the last decade that social networking services have proliferated and their use has become a widespread practice—particularly amongst young people (Johnson et al. 2009). Social media is generally used to describe collaborative media creation and sharing on a fairly large scale but is extended to include micro-communities. The uptake of social media in its various forms signifies a shift in how technology use has changed from primarily an information and entertainment source to one of communication. Young people are consuming, producing, sharing and remixing media (Burns et al. 2008). This participatory media environment enables young people to engage in creative content production, empowering them with new means of creating and sustaining connections with others. It has also opened up new debates on how to conceptualise and promote what has come to be termed cyber citizenship or digital citizenship (Coleman and Rowe 2005).

In South Africa, thousands of students have taken to social media to drive change as the recent '*fees must fall*' student protest actions in South Africa demonstrated. Social media platforms such as Facebook and Twitter buzzed with live updates of the protest, and as it gained momentum in South Africa, so did it gain even international support with messages of solidarity streaming in from institution across the globe. Social media is the voice of the youth today.

6.8 Case Study: Northern Areas Youth Leadership Academy (YLA)

This initiative was established in 2011 with the first series of activities commencing in 2012. The Northern Areas Youth Leadership Academy's goal is to educate the youth, strengthen the family and rebuild the community by aiding and supporting the educational, spiritual, moral and social development of at-risk-youth. The initiative also aims to use ICT and its various forms as an instrument to create micro-communities or clusters of young people who take ownership of the society that they will grow to inherit. It then envisioned that ICT and then more specifically mobile technology could support an increased civic engagement.

The use of mobile technology as a means of sharing information, encouraging and education at-risk youth has resulted in the establishment of an engaged and connected community. Not only are there now a new generation of leaders but also a new generation of leaders who have a sense of purpose and the will to bring about change in their society. The case study did not follow the traditional six steps of a case study as described by Yin (2014). Instead, the underlying principles of change management were employed to create the micro-community that uses technology to drive change.

6.9 Phases of Drive Change

Preparation phase

This intervention differs from other socio-economic initiatives in the Northern Areas as it is giving voice to the young people living in the community. It aims to channel youth energy in a positive way to restore the social structures in their communities. In order to do this, the initiative builds on existing capacities by strengthening natural leaders and building leadership capacity in those who exhibit leadership characteristics. School principals are requested to identify potential candidates in Grade 11. Selection of participants is based on a screening document but it is up to the individual to decide whether or not to enrol in the academy.

Manage expectations and outcomes phase

Enrolled youth are expected to commit two hours of their time every Saturday afternoon, from March till November. During this time, they participate in interactive workshops, seminars and other activities aimed at building leadership skills but which is also fun. University lecturers, community leaders and other subject experts offer their time and expertise to engage with the participants during these sessions. Recognition is given to those participants who exhibited dedication by attending a minimum of 70% of the sessions, during a formal event. This event marked the end of the programme and is attended by all involved as well the parents and guardians of the participants. Each participant is expected to share his/her experience of the programme with the audience. This in itself is a significant indicator of the level of growth of each individual.

Develop emotional quotient (EQ) phase

Emotional intelligence plays a significant role in the development of young people as it allows them to recognise their own and other people's emotions, to discriminate between the different feelings and to act upon it in an appropriate manner. EQ subject experts facilitate workshops to build EQ skills and social competence, especially for young people whose mode of communication is now via social networks and virtual communities.

Create street law awareness phase

In order to take full advantage of the changing face of communication in terms of at-risk youth, one has to understand the challenges they are faced with on a daily basis. It is therefore not only necessary to sensitise the youth about bad or potentially bad situations, whether it is emotional or physical abuse, sociological and psychological pitfalls or improper practices at institutional level. But recognition of potentially dangerous situations alone does not empower a person to act appropriately. There is thus a need to empower the youth to know their rights and to take action.

Street law taught by law students from the Faculty of Law at the NMMU has therefore become an integral part of the YLA programme. Participants simulate potentially bad situations and are guided by the facilitators on how to act upon each situation. This led to the establishment of Street Law WhatsApp groups where group members can get support or even summon help when in a bad situation. For the girls, this proved to be exceptionally valuable.

Create awareness of responsible digital citizenship phase

The perceived distance and anonymity created by electronic communication methods have resulted in unacceptable and undesirable communication behaviours among the youth. A study conducted in Nelson Mandela Bay in Port Elizabeth among 1594 primary and secondary school children, indicated that 36% of the respondents had experienced some form of cyberbullying (Badenhorst 2011). South Africa is lacking specific legislation dealing with cyberbullying leaving victims to rely on remedies offered by criminal law or civil law. Youth leaders have a significant role to play in creating awareness about the types of behaviour that constitute cyberbullying as well as how and where cyberbullying can be reported. Being part of virtual communities themselves, allow them to build positive digital citizenship among members and share the various helplines and support groups with their peers.

6.10 Using ICT to Drive Change

Understanding change

In order to effect change, one has to know what the condition is that needs to be changed, what the preferred future condition should be and how to go about it. New youth leaders are therefore put through a structured change management process to drive change, where after they have to apply it to their personal lives. This exercise provides valuable insight into the kind of undesirable situations young people are faced within their communities. Issues such as drug and alcohol abuse, gangsterism, teenage pregnancies, violence, disrespect and disorderly behaviour impact negatively on their lives. Early school dropouts are leaving the available workforce unskilled, thus adding to the already high level of unemployment.

The role of ICT

Today's youth are much more comfortable with using technology than previous generations. They are therefore encouraged to develop the capacity to solve challenges and so drive change in their own communities using mobile technology. The Northern Areas computer facilities were established with specifically this in mind. Not only are desktop computers available for computer literacy training but also tablets to make technology attractive for the younger generation. At-risk youth are encouraged to make use of the facility to first educate them about ICT facilitates the creation and sharing of information. The focus is on protecting their will to use ICT to drive change in a manner that will encourage change but keeping the potential threats in mind. Threats include the risk of cybercrime and bullying, inappropriate use of the technology to support criminal activities or even the use of technology for the production of inappropriate materials such as pornography.

Since 2012 to date, the use of various forms of technologies is observed. The technologies range from structured uses in a controlled environment such as the computer facilities in a protected area to own mobile phones in a virtual world with potentially an unknown identity. Over time, the following lessons and changes were observed:

- **Individual identity and self-expression** amongst the youth of the Northern Areas is developing at an increasing rate. The flexibility of mobile technology to encourage individual customization also enhances the ability to express identity. Mobile technology is found to through the use of social media reinforce parts of youth identity including societal and cultural background.
- **Strengthening of interpersonal relationships** through the use of instant messaging, WhatsApp and social networking. These technologies address the specific barriers in the Northern Areas that young people may face to forming and maintaining positive social relationships. Mobile technology has removed these barriers because they are accessible 24/7, from different physical locations and via different technologies such as mobile phones and tablet PCs. The use of different forms of online social networking, such as instant messaging, is now established to support networks with peers. In addition, youths who are marginalised and socially isolated are now using the opportunity to socialise using technology.
- **Convergence of online and offline spaces** is now on the increase. It is evident in the observations that young people are increasingly engaging simultaneously in online and offline social networking. It appears that young people often work collaboratively in the online space through WhatsApp, for example, creating or commenting on YouTube videos or other such activities, while physically co-located. They not only consider their online and offline worlds as one but also use this to create a sense of belonging and connectedness with peers.
- **Civic engagement via technology** has evolved into an opportunity for the Northern Areas youth to share information and to create issue-oriented groups. Social networking services are used to find out what other people are doing by connecting with individuals with similar interests, existing activities or disseminating information about their own projects. At-risk youth are now creating new participatory communities by and for their peers to allow participation in a virtual environment rather than facing the day-to-day challenges of the Northern Areas.

6.11 Humanising Pedagogy

For decades, educators and others considered various strategies to close the gap for at-risk youth. They have sought solutions involving new uses of technology. In some instances, the results of technology initiatives have been mixed. Often the introduction of technology has failed to meet the expectations leaving specifically the educational landscape replete with stories of how at-risk youth were unable to benefit from particular innovations. In this particular initiative, an approach of humanising pedagogy was taken. Rather than using technology to drill and kill the enthusiasm of the youth, technology was used as an interactive instrument. Through the use of interactive engagement using technology, the notion of vulnerability is explored and employed as a developmental and human security context that explores the interdependence between human beings and technology. Ultimately, the aim is to develop an embedded pedagogy that can enhance educational engagement in a non-threatening manner using technology.

6.12 Conclusion and Future Work

The use of mobile technologies and then specifically the instant messaging and social media are transforming communication practices, opening new spaces and processes of socialisation and impacting upon traditional social structures. These effects are particularly relevant for the most frequent users of the technologies namely youth. This new environment poses certain challenges and, like any setting for social interaction, has some inherent risks. However, this paper suggests that through increased Internet and media literacy—ensuring all young people develop the skills to critically understand, analyse and create media content—these challenges can be overcome and risks mitigated in a way that ensures the many benefits of ICT and more specifically mobile technologies can be realised.

In other words, by maximising the benefits of mobile technology, whether it is its role in delivering educational outcomes, or facilitating supportive relationships, identity formation, or a sense of belonging and resiliency, many of the risks of virtual interaction, such as cyberbullying, can be minimised. Strategies to this end must be underpinned by best practice change management evidence and more research should be undertaken to ensure that emerging practices and effects of mobile technology as a driver of change are understood and responded to.

This paper finds that the benefits of mobile technologies are largely associated with the participatory nature of the modern digital environment. There remains a need for exploratory academic work on cybercitizenship amongst at-risk youth and the ways in which such youths are engaging online to express their views, challenge and create their own modern views on society. This knowledge could lead to the recognition that their online practices may challenge commonly held notions about childhood, youth, gender and other critical social issues.

It is important to note that new digital communication practices could well be informed by the experiences and perspectives of the new generation of mobile youth with their view of digital literacy. This in turn could lead to a better understanding of how the online and offline worlds are integrated to drive change whilst using ICT as the tool for change.

Ongoing evaluation of the intervention feeds into the offerings and influence the respective technologies and content that is contained in the intervention. The next step is to now replicate the intervention in similar communities based on the lessons learnt and the good practices that have evolved from this exploratory study. Some examples for replication include the use of similar technologies for at-risk groups including refugees, abused women and as a means to support addiction treatment. Some considerations for the replication of the study would include connectivity, appropriate support applications that allow for the different types of support required, and appropriate interaction techniques to allow for varying digital literacy levels.

References

Badenhorst, C. (2011). Legal responses to cyber bullying and texting in South Africa: Centre for Justice and Crime Prevention CJCP Issue Paper No 10.

Burns, J. M., Durkin, L. A., & Nicholas, J. (2008). ReachOut! *The Internet as a setting for mental health promotion and prevention in Éisteach: Journal of the Irish Association of Counseling and Psychotherapy, 8*(1), 13–19.

Coleman, S., & Rowe, C. (2005). Remixing citizenship: Democracy and young people's use of the internet a report for the Carnegie Young People's Initiative.

Eyebers, S and Giannakopoulos, A. (2015) The Utilisation of Mobile Tecnologies in Higher Education: Lifebuyo or Constriction? In mLearn 2015 CCIS 560, pp. 300 – 315. T.H. Brown and H.J. van der Merwe (Eds) in Springer Switzerland.

Johnson, L., Levine, A., & Smith, R. (2009). *The 2009 horizon report*. Austin, Texas: The New Media Consortium.

Kaiser Family Foundation. (2007). Young South Africans, broadcast media, and HIV/AIDS awareness: Results of a national survey, 2007, URL http://www.kff.org/southafrica/upload/7614pdf. Accessed January 22, 2016).

Matavire, M. (2007, May 19). Bhisho has reneged on promise, says ANC. *The Herald*, 9.

Statistics South Africa (StatsSA). (2015). Sensus 2011. http://www.statssa.gov.za/publications/P03014/P030142011.pdf. Accessed January 22, 2016).

Third and Richardson. (2010). *Connecting, supporting and empowering young people living with chronic illness and disability: The livewire online community*, (Report prepared for the Starlight Children's Foundation, January 2010). SBN: 978-0-86905-997-5.

Yin, R. K. (2014). *Case study research. Design and methods* (5th ed.). Thousand Oaks: Sage Publications.

Part III
Medical Technologies

Chapter 7
Barriers to Point of Care Testing in India and South Africa

Nora Engel, Vijayashree Yellappa, Malika Davids, Keertan Dheda, Nitika Pant Pai and Madhukar Pai

7.1 Introduction

Point of care (POC) testing in communities, home settings, and primary healthcare centers is widely believed by the global health community to have tremendous potential in reducing delays in diagnosing and initiating treatment for diseases such as HIV, tuberculosis, syphilis, and malaria. The idea is that testing nearer to the patient, at the point of care, allows for quick diagnosis and further management decisions (referral, follow-up testing or treatment) completed in the same clinical encounter or at least the same day, while the patient waits. In this way, the POC continuum is ensured (Pant Pai et al. 2012). POC testing promises to overcome long turnaround times and delays associated with conventional laboratory-based testing. These problems can result in the loss of patients from testing and treatment pathways with detrimental consequences for the development of advanced disease stages and drug resistance.

N. Engel (✉)
Maastricht University, Maastricht, Netherlands
e-mail: n.engel@maastrichtuniversity.nl

V. Yellappa
Institute of Public Health, Bangalore, India
e-mail: vijayashree@iphindia.org

M. Davids · K. Dheda
University of Cape Town, Cape Town, South Africa
e-mail: malika.davids@uct.ac.za

K. Dheda
e-mail: keertan.dheda@uct.ac.za

N. P. Pai · M. Pai
McGill University, Montreal, Canada
e-mail: nitika.pai@mcgill.ca

M. Pai
e-mail: madhukar.pai@mcgill.ca

S. Hostettler et al. (eds.), *Technologies for Development*,
https://doi.org/10.1007/978-3-319-91068-0_7

The devices that lend themselves to such POC testing are often thought to be simple, cheap, and rapid (Mabey et al. 2012), work without access to laboratories, fridges, gloves, biosafety, continuous power supply, or trained staff and meet what the World Health Organization defined as ASSURED criteria (Affordable, sensitive, specific, user-friendly, rapid and robust, equipment free, and delivered). However, the availability of relatively cheap, simple, and rapid tests that can be conducted outside laboratories does not automatically ensure the POC continuum. We know that f.i. malaria rapid tests are often not used or the results not acted upon (Chandler et al. 2012). Similarly, TB tests deployed at POC might require additional infrastructural, financial, and operational support, exhausting resources at the clinic (Clouse et al. 2012), and HIV rapid testing is at times hampered by poor linkages to care (Kranzer et al. 2010). In order to understand the new roles and challenges that medical devices such as POC tests encounter, we need to study diagnostic practices at the POC and how devices are integrated into workflow and patient pathways.

This chapter reviews results from a qualitative research project on barriers to POC testing in South Africa and India. In this project, we aimed at understanding where POC testing is happening and what the main barriers are. Using a framework that envisions POC testing as programs, rather than just tests, across five settings (home, community, peripheral laboratory, clinic, and hospital) (Pant Pai et al. 2012), we examined diagnostic practices across major diseases and actors in homes, clinics, communities, hospitals, and laboratories in South Africa and India. Detailed results per country have been published (Engel et al. 2015a, b, c). Here, we review selected results, discuss them comparatively, and reflect on the implications for medical device design.

7.1.1 Qualitative Project on Barriers to POC Testing

Data for this project was collected in semi-structured interviews (N = 101 in South Africa, N = 8 in India) and focus group discussions (N = 7 in South Africa, N = 13 in India) with doctors, nurses, community health workers, patients, laboratory technicians, policymakers, hospital managers, and diagnostic manufacturers between September 2012 and June 2013 in Durban, Cape Town, and Eastern Cape (South Africa) and Bangalore and Tumkur district (India). Participants were purposively sampled to represent the settings of hospitals, peripheral labs, clinics, communities, and homes in both the public/private sector and rural/urban setting. In the context of conducting interviews and FGDs, we visited labs, clinics, and testing facilities. These three data sources allowed us to triangulate data. The interviews specifically examined diagnostic steps for each major disease occurring in the setting (such as HIV, TB, diabetes, diarrhoeal diseases and hypertension in South Africa and HIV, TB, malaria, hepatitis, syphilis, diabetes, typhoid, and dengue in India) in great detail from ordering a test to acting on a result, including available material and capacities, turnaround times, and referral processes. Additionally, we explored during interviews the challenges that participants encountered when diagnosing, understanding of diagnosis, and visions of an ideal test. The focus group discussions focused exclusively on challenges experienced when diagnosing. Interview and focus group discussion guides were piloted and revised during the fieldwork to improve the clarity of questions. All interviews and discussions were held in English (except some of them in Kannada in India) and digitally recorded, and the notetaker wrote down main points raised, nonverbal communication, and setting characteristics. Data analysis was done thematically (Eisenhardt and Graebner 2007), using Nvivo 9 (QSR International).

7.1.2 Ethics Approval

The ethics review board of the University of Cape Town, South Africa and McGill University Health Centre (MUHC), Montreal, Canada approved this study. Approvals for interviews and discussions conducted at public primary healthcare facilities were sought from the Provincial Department of Health authorities as necessary. All participants were provided with information sheets explaining the objectives of the study and all signed informed consent forms prior to participation.

7.2 Results

In the following, we review selected results per country. This highlights the very different diagnostic eco-systems (see Figs. 7.1 and 7.2) and the distinct set of challenges to diagnosing at POC, as well as the strategies actors employ to overcome these.

7.2.1 India

India has a highly fragmented and largely unregulated diagnostic landscape. Laboratory-based testing takes place across a multitude of providers ranging from small, ill-equipped one-room laboratories in public clinics to large hospital laboratories, from small private neighborhood laboratories with limited testing equipment to medium-sized facilities and state-of-the-art laboratory chains. Patients carry the main responsibility for ensuring a POC continuum. Patients in India are the carriers of samples, reports, and communication between the providers. If they are asked to obtain a diagnostic test in any of the settings, they need to go to the laboratory themselves, provide a sample there, pick up results once available, and return them to the doctor. The system thus relies heavily on patients' initiative to ensure successful POC testing. Oftentimes, these journeys start in the private sector (for rich and poor patients alike) and if patients are not diagnosed and treated successfully, these

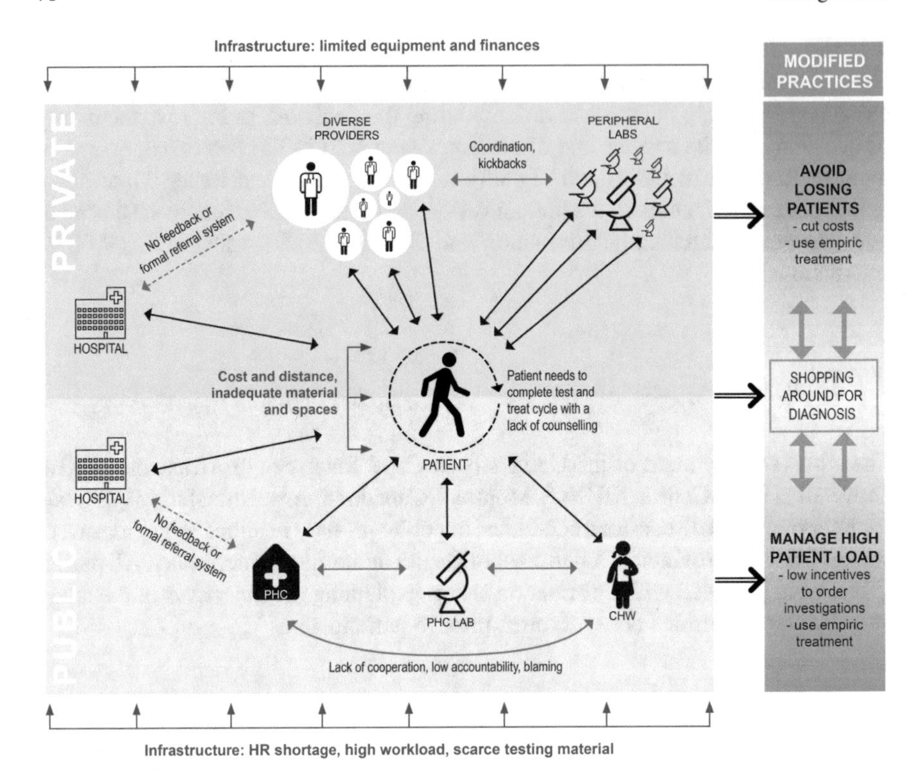

Fig. 7.1 Diagnostic pathway and challenges to POC testing in India (figure first published in Engel et al. 2015b)

journeys can be long, frustrating, expensive, and exhausting or confusing and then chances are high that patients opt out.

Most private practitioners do not conduct any tests on the spot but send their patients to small peripheral laboratories nearby, with whom they have aligned their opening hours and established so-called kickbacks for referring patients (laboratories pay practitioners 30–40% of the price for each test ordered). These small peripheral laboratories offer a variety of tests including HB card method, random blood sugar (glucometer or manual), urine dipstick or chemical analyser, platelets, complete blood count (CBC), malaria using microscope (rarely rapid card test), blood grouping, HBsAG (hepatitis), VDRL card test (syphilis), typhoid slide (Widal), TB Mantoux, and HIV Tridot. They usually cannot afford rapid test kits and their reagents. But due to smaller volumes (10–30 patients a day), these laboratories are able to maintain a one-hour turnaround time using older methods (Engel et al. 2015c). We showed that thanks to this coordination, a patient seeing a doctor in the morning, can get tested by a nearby lab and return the results to the treating doctor in the afternoon or evening. The coordination thus ensures a POC continuum. Yet, kickbacks can also cause malpractice and a patient's knowledge of these arrangements can lead to

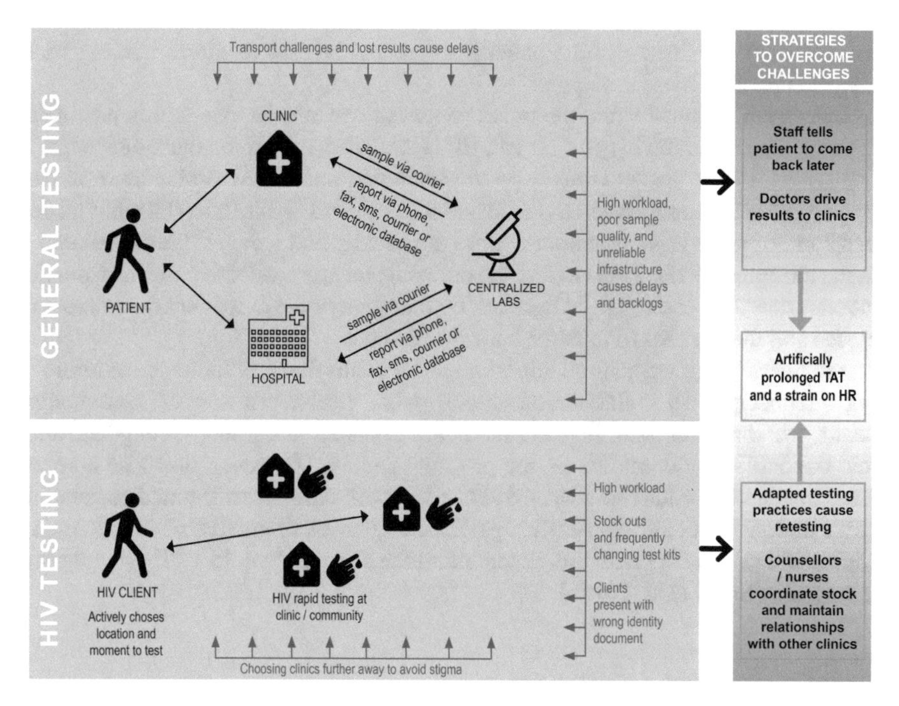

Fig. 7.2 Diagnostic pathway and challenges to POC testing in South Africa

distrust into the care provided, causing patients to change providers and seek care elsewhere (Engel et al. 2015b).

In the public sector, primary public health centers are usually equipped with a small laboratory staffed with a laboratory technician who can run basic tests, such as malaria smears, HbsAg card test (hepatitis), HIV rapid and Coomb's tests, dengue NS1 card test and dengue through IgG and IgM lab tests, urine dipstick, urine sugar testing with Benedict's solution, urine albumin, and total and differential count of white blood cells. However, these laboratories often suffer from underfunding, weak infrastructure, limited budgets for reagents and test kits, and high workloads leading to delays in turnaround times (Engel et al. 2015c). This means that patients are told to come back the next day for results of basic investigations or that they need to be referred for testing to the (sub)-district hospital further away. It also means that medical officers are more inclined to start treating empirically than based on diagnostic test results. Medical officers in public health centers can see up to 100–150 patients per day and lack the time to order investigations or are hesitant due to expected delays or absent materials in the laboratory. If patients accessing these centers are told to come back the next day for results or are referred to another hospital, cost for testing (including potential user fees, transport (from home, to clinic, to diagnostic centers), food and/or accommodation, drugs, and often loss of daily wages) increases and the patients might not be able to afford coming back (Engel et al. 2015b).

Loss from diagnostic and treatment pathways can also happen within hospitals. Patients who are asked to get a TB and HIV test done in a hospital outpatient department, for instance, need to queue, provide samples, and return to pick up results in two different laboratories across the hospital compound. While hospital laboratories make use of rapid tests, turnaround times are usually half a day. If the patient came in the morning, he/she needs to find the doctor again after collecting results from the laboratories. The doctor might have left by that time or the outpatient department is closed and they are asked to come back the next day.

Providers suggested centralizing laboratories in hospitals compounds and improving interdisciplinary collaboration among public health center staff. Community health workers suggested that testing at the doorstep could help strengthen trust into the public health clinics among potential patients. However, they also emphasized that testing would have to be followed up with basic treatment and counseling at the doorstep too, since absent drugs, follow-up tests, or doctors at the clinics, as well as irregularly supplied tests to conduct at the doorstep, would spoil that renewed trust again (Engel et al. 2015b).

7.2.2 South Africa

Very different from India, South Africa has a highly centralized diagnostic landscape. Diagnostic testing in the public sector is provided by the National Health Laboratory Service (NHLS), while a handful of large diagnostic companies provide diagnostic services to private providers. This means that clinics and smaller hospitals send the majority of their samples via courier to a centralized laboratory, usually in one of the bigger cities, and results are sent back by courier, Internet, SMS, phone, or fax. A few tests are conducted on the spot in public and private clinics. Among them are basic screening tests (blood pressure, weight, glucose, rapid HB, and urine dipstick) and the HIV rapid tests to establish whether someone is HIV infected. In selected public clinics, the Xpert MTB/RIF, a molecular test that promises to diagnose tuberculosis in 90 min, is being implemented [while the majority of the Xpert instruments have been deployed in centralized NHLS laboratories (Cohen et al. 2014)]. For those tests conducted on the spot, results are available within the same patient encounter. An exception is Xpert MTB/RIF where large numbers of samples lead to backlogs and increase the turnaround time from 90 min to 24 h (since only four samples can be run in parallel) (Engel et al. 2015a); in addition, it requires 2,5 extra staff to operate the device (Clouse et al. 2012). Treatment initiation can again lead to delays. In the case of HIV, follow-up testing and counseling sessions mean that treatment is initiated 4–7 days after a positive result was established in clinics (Engel et al. 2015a).

The providers reported delays for those diagnostics that are run in centralized laboratories and that no same-day results were available for those tests. Mainly due to transportation challenges (long distances, poor roads, strikes, bad weather, shortages of couriers, and damaged samples), but also due to poor sample quality, lost results, misspelled names, excess workload in laboratories, and break down of computer systems. In anticipation of these delays, doctors and nurses would tell patients to come back even later, would drive results out to clinics by themselves, or fetch results in laboratories. These strategies artificially prolong diagnostic turnaround times and put a strain on scarce available human resources (Engel et al. 2015a).

Counselors and nurses who conduct HIV rapid tests (lateral flow assays using blood samples generated by a finger prick) in clinics, outreach settings, and communities need to maintain relationships to other providers to help each other in overcoming stockouts of test kits and reagents and manage high workloads. This could mean that counselors and nurses would share the number of people who came for testing or that hospital doctors would drive test kits to remote clinics based on a good relationship with the nurse working in that clinic (Engel et al. 2017). Irregular supply of test kits, frequently changing test kit brands with slightly different steps to perform (for instance time to result, amount of buffer added), or efforts to accommodate impatient clients mean that providers described different ways of using the test kits (for instance, how long they waited before reading results of the testing strip). These adaptations in practices and differences in how HIV rapid tests were conducted have unclear implications for outcomes of testing and mean that some providers mistrusted each other's quality of diagnostic work, leading to replication of testing and additional delays (Engel et al. 2017). Some providers suggested to establish comprehensive quality control measures, systems of maintenance, and ensure continuous (re)-training for every new rapid test kit that counselors and nurses use. While some wished more testing could be done at POC to guide on the spot treatment decisions or at doorsteps in communities to reduce fear of social stigma when accessing clinics, others emphasized to instead prioritize improving referral systems to centralized laboratories, staff training, and infrastructure strengthening.

Clients who seek an HIV test actively manage their diagnostic processes including attempts to control where they receive testing (by choosing testing sites that fit considerations of cost, distance, or avoidance of social stigma), how and which results they are able to obtain. This could sometimes mean to present with a wrong identity to be eligible for testing at a different site or to access a different diagnostic device (in this case, POC CD4 testing, an HIV monitoring test that was only available in selected sites to newly diagnosed clients). These strategies burden an already overstretched public health system by using additional testing resources and creating difficulties with follow-up and tracing. However, they also show how patients actively turn the test into a tool to achieve another goal (in this case to access another testing device) (Engel et al. 2017).

7.3 Discussion and Conclusion

Our results show how the specific diagnostic eco-systems of South Africa (largely centralized testing with the prominent exception of HIV screening) and India (largely peripheral testing spread across a multitude of providers) constitute very different conditions for POC testing and how the major challenges to ensuring POC continuums are linked to this difference.

In India, successful POC testing hardly occurs in any of the settings and for any of the diseases. Many of the rapid tests are used in laboratories where either the single patient encounter advantage is not realized or the rapidity is compromised due to human resources, manpower, and equipment shortages. In smaller peripheral private laboratories and private clinics with shorter turnaround times, rapid tests are unavailable or too costly. The onus to follow through diagnostic pathways is on the patients and providers use coordination mechanisms (opening hours, kickbacks) to ensure some form of POC continuum using older technologies. In South Africa, the majority of testing happens in centralized laboratories, where delays are accumulated due to transportation, human resource, and infrastructure challenges. Providers' strategies to deal with associated delays create new problems, such as artificially prolonged turnaround times, strains on human resources, and quality of testing, compounding additional diagnostic and treatment delays. While most tests conducted on the spot can be made to work successfully as POC tests, delays remain with regard to treatment initiation.

In both countries, actors use different strategies to overcome these challenges. These adaptive strategies to make POC testing work are in both countries rather fragile and ad hoc, dependent on locally negotiated solutions, personal commitment, available human resources, and relationships. While some are successful in ensuring a timely diagnosis, others lead to disruptions, unnecessary testing, or delays with at times unclear implications for quality of diagnosis. In India, strategies for coordinating between private providers and laboratories with kickbacks or treating empirically right away are aimed at avoiding losing patients, and in this way ensure some form of POC continuum. However, they also increase chances for malpractices (wrong or unnecessary tests might be ordered, inadequate treatment might be prescribed) and may lead to mistrust from patients into the health system. In South Africa, strategies of dealing with delays associated with centralized testing actually increase diagnostic delays (such as telling patients to come back even later) or put additional strains on the health system (such as doctors delivering results to remote clinics). Testing on the spot, for instance, HIV testing, requires healthcare providers to maintain functioning relationships to other providers to overcome stockouts and excess workload, while their adaptations in conducting HIV rapid tests to continuously changing test kits or patient demands can foster mistrust among providers.

Patients embody very active roles in managing their diagnostic journeys. In India, the system relies almost entirely on the patient to ensure the POC continuum across homes, clinics, labs, and hospitals, amidst a multitude of public and private providers with divergent and often competing practices in settings lacking material, money, and human resources. The onus is on the patient to ensure completion of test and treat cycles. If a patient's initiative is not supported in these journeys, chances are high that he/she opts out. Constructive counseling by providers about various aspects of diagnostic tests and processes is necessary but not sufficient. Functioning relationships between providers are equally important (Engel et al. 2015b).

While the system in South Africa does not foresee such an active role for patients in ensuring a POC continuum, it still relies on a patients' ability to return to clinics on another day for either results (in the case of laboratory-based testing) or for follow-up testing and counseling sessions in the case of HIV. The examples of HIV testing showed that HIV patients also need to actively manage diagnostic processes to make testing fit their personal circumstances and make testing worthwhile for themselves (Engel et al. 2017).

These profound differences between the diagnostic setup, offer very different conditions for POC testing. Although the promises that have been attached to POC testing easily lend themselves to view testing technologies as silver bullets, it matters how diagnostic processes are organized and made to work at POC. The results reveal how the material dimensions of diagnosis, such as the test platform, reagents, and supplies, the actors involved, their relations and the sociocultural context in which testing and diagnosis are happening are invariably interlinked (Engel et al. 2015b). This means that simply focusing on one element, for instance, improving infrastructure or test platforms or relationships, is not sufficient. Those aspects need to be studied and tackled together. The contrasting results from India and South Africa further highlight that the settings and the tests have their own histories, assumptions, practices, and understandings inscribed in them. This means that by implementing tests successfully, both the setting, including its organization of the workflow, workforce, its infrastructure, interaction with patients, and standards, and the tests are being shaped and need to be adapted. Tools and (user) practices are being co-constructed (Oudshoorn and Pinch 2003). Implementing diagnostic tests is thus a dynamic and ongoing process that requires continuous observation, analysis, reflection, iterations, and adaptations of tests and (user)-practices and -settings.

Such insights need to be taken into account when designing POC testing programs and technologies. Test developers, decision-makers, and funders need to account for these ground realities. They need to identify and involve users and various stakeholders in design, evaluation, and implementation processes. However, current global diagnostic design and development practices, research, regulation, and evaluation capacities do not do justice to the dynamic nature of these processes of making diagnostic tests work at POC.

Based on these insights, practitioners, donors, and test developers should make sure to

- Study diagnostic practices at POC and how devices are integrated into workflow and patient pathways before, during and after design and implementation of new products;

- Tackle jointly the relationships between providers and between patients and providers, infrastructure, testing platforms, and adaptive strategies of dealing with constraints;
- Examine the role, responsibility, and work of patients to ensure a POC continuum;
- Allow continuous observation, analysis, reflection, iterations, and adaptations of tests and (user)-practices and -settings; and
- Develop research capacity to assess and integrate these factors.

References

Chandler, C. I., Mangham, L., Njei, A. N., Achonduh, O., Mbacham, W. F., & Wiseman, V. (2012). As a clinician, you are not managing lab results, you are managing the patient: How the enactment of malaria at health facilities in Cameroon compares with new WHO guidelines for the use of malaria tests. *Social Science and Medicine, 74,* 1528–1535. https://doi.org/10.1016/j.socscimed.2012.01.025.

Clouse, K., et al. (2012). Implementation of Xpert MTB/RIF for routine point-of-care diagnosis of tuberculosis at the primary care level. *South African Medical Journal, 102,* 805–807.

Cohen, G. M., Drain, P. K., Noubary, F., Cloete, C., & Bassett, I. V. (2014). Diagnostic delays and clinical decision-making with centralized Xpert MTB/RIF testing in Durban, South Africa. *JAIDS Journal of Acquired Immune Deficiency Syndromes, 67,* e88–93. https://doi.org/10.1097/QAI.0000000000000309.

Eisenhardt, K. M., & Graebner, M. E. (2007). Theory building from cases: Opportunities and challenges. *Academy of Management Journal, 50,* 25–32.

Engel, N., Davids, M., Blankvoort, N., Dheda, K., Pant Pai, N., & Pai, M. (2017). Making HIV testing work at the point of care in South Africa: A qualitative study of diagnostic practices. *BMC Health Services Research, 17,* 408. https://doi.org/10.1186/s12913-017-2353-6.

Engel, N., Davids, M., Blankvoort, N., Pai, N. P., Dheda, K., & Pai, M. (2015a). Compounding diagnostic delays: A qualitative study of point-of-care testing in South Africa. *Tropical Medicine and International Health, 20,* 493–500. https://doi.org/10.1111/tmi.12450.

Engel, N., Ganesh, G., Patil, M., Yellappa, V., Pant Pai, N., Vadnais, C., et al. (2015b). Barriers to Point-of-Care testing in India: Results from qualitative research across different settings, users and major diseases. *PLoS ONE, 10,* e0135112. https://doi.org/10.1371/journal.pone.0135112.

Engel, N., Ganesh, G., Patil, M., Yellappa, V., Vadnais, C., Pai, N., et al. (2015c). Point-of-care testing in India: missed opportunities to realize the true potential of point-of-care testing programs. *BMC Health Services Research, 15,* 550.

Kranzer, K., et al. (2010). Linkage to HIV care and antiretroviral therapy in Cape Town, South Africa. *PLoS ONE, 5,* e13801. https://doi.org/10.1371/journal.pone.0013801.

Mabey, D. C., et al. (2012). Point-of-Care tests to strengthen health systems and save newborn lives: The case of syphilis. *PLoS Med, 9,* e1001233. https://doi.org/10.1371/journal.pmed.1001233.

Oudshoorn, N., & Pinch, T. J. (2003). Users and Non-Users Matter. In N. Oudshoorn & T. J. Pinch (Eds.), *How users matter: The co-construction of users and technology* (pp. 1–25). London, Cambridge MA: MIT Press.

Pant Pai, N., Vadnais, C., Denkinger, C., Engel, N., & Pai, M. (2012). Point-of-Care testing for infectious diseases: diversity, complexity, and barriers in low- and middle-income countries. *PLoS Medicine, 9,* e1001306. https://doi.org/10.1371/journal.pmed.1001306.

Chapter 8
Health Hackathons Drive Affordable Medical Technology Innovation Through Community Engagement

Aikaterini Mantzavinou, Bryan J. Ranger, Smitha Gudapakkam, Katharine G. Broach Hutchins, Elizabeth Bailey and Kristian R. Olson

8.1 Introduction

Successful medical innovation requires a process of co-creation among key healthcare stakeholders including healthcare professionals, end users, scientists, engineers, and entrepreneurs (IDEO 2015; Lee et al. 2012; Prahalad and Ramaswamy 2004). This can be challenging especially in resource-limited settings where interdisciplinary collaboration may be hampered by more pronounced professional, socioeconomic, and age barriers (Sachs 2003).

MIT Hacking Medicine, a group founded in 2011 at MIT, aims to energize the healthcare community and accelerate medical innovation by carrying out co-creation through health hackathons. These 1- to 3-day events bring together diverse stakeholders to solve pressing healthcare needs. The group has organized to date more than 40 health hackathons across 9 countries and 5 continents.

Aikaterini Mantzavinou and Bryan J. Ranger are co-first authors.

A. Mantzavinou (✉) · B. J. Ranger (✉)
Harvard-MIT Program in Health Sciences and Technology, Cambridge, MA, USA
e-mail: amantzav@mit.edu

B. J. Ranger
e-mail: branger@mit.edu

S. Gudapakkam · K. G. Broach Hutchins · E. Bailey · K. R. Olson
Consortium for Affordable Medical Technologies, Boston, MA, USA
e-mail: sgudapakkam1@babson.edu

K. G. Broach Hutchins
e-mail: kghutchins@partners.org

E. Bailey
e-mail: baileyelizabetha@gmail.com

K. R. Olson
e-mail: krolson@mgh.harvard.edu

S. Hostettler et al. (eds.), *Technologies for Development*,
https://doi.org/10.1007/978-3-319-91068-0_8

MIT Hacking Medicine first partnered with the Consortium for Affordable Medical Technologies (CAMTech) of the Center for Global Health at the Massachusetts General Hospital (MGH) in the fall of 2012 to organize a hackathon focused on affordable medical technology for low- and middle-income countries (LMICs). Since then, MIT Hacking Medicine's annual flagship event has featured a global health track in collaboration with CAMTech whose participants include students and healthcare professionals from India and Uganda. The Hacking Medicine team has in turn joined CAMTech-led hackathons in India and Uganda. Engagement of the local community in the process of healthcare disruption has led to multiple award-winning projects for better monitoring, diagnosis, treatment, and prevention of disease. Many of these solutions are translated from bench to bedside at remarkable rates aided by the CAMTech Innovation Platform and follow-up strategies that promote and track project development and progress. Sustainable business strategies for product commercialization and reverse innovation to bring cost-effective technologies to resource-rich settings are highly encouraged. The hacking philosophy applied to affordable healthcare, engagement of the local community engagement, and support of novel ideas toward long-term sustainability hold great promise for creating low-cost medical solutions that can improve healthcare outcomes globally.

8.2 Objectives and Methodology

8.2.1 The Need for Co-creation

Both developed and developing economies are under pressure to provide more cost-effective healthcare to address the demands of their populations (Chilukuri et al. 2010; Witty 2011). Medical technology companies are rethinking their approach to medical product design to increase product efficiency and accessibility (Prahalad and Ramaswamy 2004; Witty 2011). This is particularly true in resource-limited areas, where political, financial, and cultural constraints often hamper innovation and stakeholder collaboration necessary for collective healthcare transformation (Chilukuri et al. 2010; Sachs 2003; Mauser et al. 2013).

8.2.2 The Need for Health Hackathons: MIT Hacking Medicine and the Hackathon Model

Locally driven medical transformation in LMICs requires crosstalk between the main players of the healthcare ecosystem. These players need to overcome communication and collaboration barriers, see beyond their discipline and approach a problem using a nontraditional way. This allows them to generate effective and efficient healthcare solutions. Health hackathons promise to offer precisely the opportunity to do so.

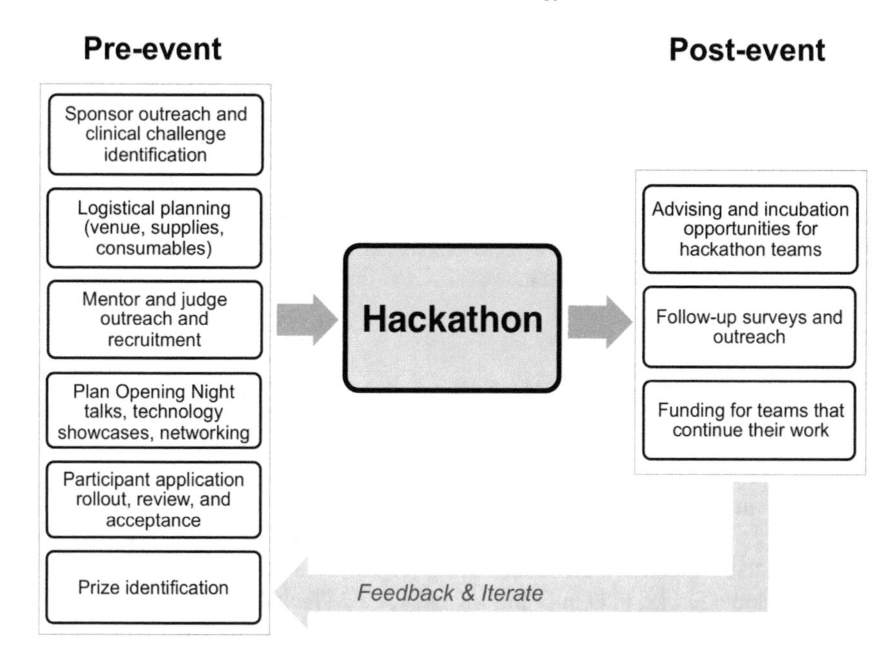

Fig. 8.1 The MIT hacking medicine health hackathon process

MIT Hacking Medicine organizes health hackathons with specific themes attended by participants with diverse backgrounds, who will address pain points in healthcare delivery and assess business viability as a key component of their hacks (Caputo 2014) (Fig. 8.1). The group has organized more than 40 hackathons across 9 countries and 5 continents. Themes have included reproductive, maternal, newborn, and child health; diabetes; telehealth; Ebola; and road safety. Sponsorships or collaborations with different entities are established for each hackathon. The hackathon scope and intended outcomes are refined closer to the event date, and expert mentors are recruited to support the participants during the hackathon. Judges are enlisted among the sponsors and healthcare innovator community to determine the hackathon winners. Participants are selected through a rigorous application that ensures diversity and interest in healthcare innovation.

MIT Hacking Medicine has focused much of its attention on creating a health hackathon experience that does not start and end at the hackathon weekend; rather, the vision is to generate a network of individuals compelled to make healthcare better by exchanging ideas, knowledge, and skills in the long term. For that reason, the group has piloted pre- and post-hackathon opportunities for networking, ideation, securing funds, and progress follow-up.

The actual hackathon begins with healthcare experts presenting theme-specific umbrella challenges. Participants then pitch project challenges to their peers focusing on a specific "pain point." These 60-second pitches energize the audience and serve as starting points for team formation and problem identification. A mingling session ensues to allow team formation. Teams spend the subsequent event duration "hacking" their projects, receiving guidance from mentors with expertise related to health care, engineering, business development, and other key areas.

At the conclusion of the event, the hackathon teams present their work to the judging panel and the rest of the participants. Presentations are limited to a 3-minute pitch capturing the clinical need, ideas for a solution, any prototypes created during the hackathon, and a business model. Prizes may include monetary awards, sponsored internships, support from startup incubators, and funds for pilot studies.

8.2.3 The MIT Hacking Medicine Model Applied to Hackathons in LMICs by CAMTech

CAMTech first partnered with MIT Hacking Medicine in 2012 to organize a hackathon in Boston focused on affordable medical technology for LMICs. The collaboration extended to a global health track sponsored by CAMTech in the 2014 and 2015 Grand Hacks, held in Cambridge MA. CAMTech further adapted the MIT Hacking Medicine health hackathon model to the LMIC health needs and innovation potential by bringing health hackathons to India and Uganda. In collaboration with MIT Hacking Medicine, CAMTech has organized eight international events in these two countries over the past 3 years. Participants and mentors from the India and Uganda ecosystems created through these local events have then joined their US-based counterparts in the 2014 and 2015 Grand Hacks, giving a voice to the LMIC setting in the Cambridge-based events. CAMTech has also organized Boston hackathons with a focus on pressing global health challenges in partnership with MIT Hacking Medicine, including a Stop Ebola hackathon held in 2015 at MGH and a Global Cancer Innovation hackathon held in 2016 at MGH.

Through this relationship, the authors have found that the hackathon model can be adapted to an LMIC setting in a straightforward manner, and presents a unique opportunity to involve local stakeholders to work toward the betterment of their community. More so than in the resource-rich parts of the world, in LMICs, physicians may never brush shoulders with software engineers, and business people may never cross paths with technology designers. A local health hackathon puts these unlikely players together at the same drawing board, encouraging different perspectives, experiences, and expertises to play off each other and so championing locally driven, sustainable healthcare improvements in a groundbreaking way.

8.3 Potential for Development Impact

Solutions coming from healthcare stakeholders in LMICs inspire entrepreneurship and confidence in the community (Morel et al. 2005a, b). By involving key figures

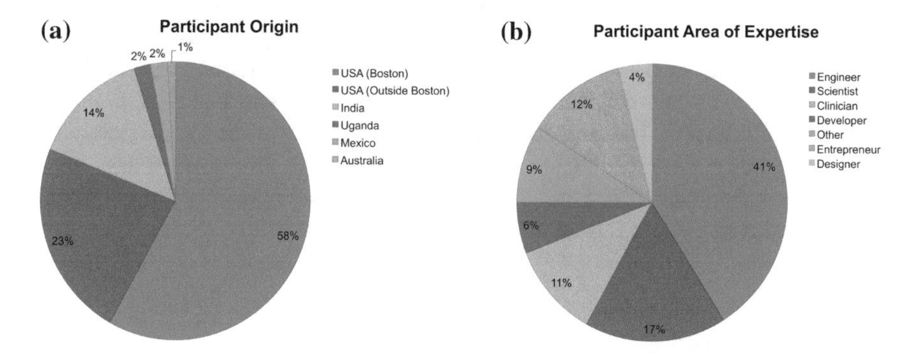

Fig. 8.2 MIT hacking medicine's grand hack 2015 statistics for global health track participants: **a** Origin of the participants; **b** Area of expertise or background

of the local healthcare delivery and technology development chains in the medical innovation process, a health hackathon in LMICs can harness local talent and knowledge to result in solutions championed by their local creators that can more easily be adapted by the community. At the same time, power holders in LMICs such as investors, entrepreneurs, educational institutions, or even governments take notice of the innovation coming out of these local hackathons, oftentimes opting to support this work financially or otherwise (Bailey 2014).

8.3.1 Direct Impact of Global Health Hackathons

Over the past 4 years, CAMTech and MIT Hacking Medicine have collaborated to run 11 global health-focused hackathons with a total of 2077 participants and over 470 pitches. The events have turned into annual initiatives in both India and Uganda, attracting several hundred participants from all over the host country and even neighboring countries. The applicant pool has grown making participant admission highly competitive. Figure 8.2 provides a profile of global health track participants at MIT Hacking Medicine's Grand Hack in 2015.

8.3.2 CAMTech Extension of the Hackathon Model

Though hackathons are continually growing in popularity on a global scale, they should not be viewed as a single event but rather a launching pad to spur sustainable innovations. CAMTech has worked to extend the hackathon model to engage stakeholders after the event, provide resources, and educate on best practices for cost-effective medical innovation. To incentivize teams to continue working on their ideas beyond the event, a post-hackathon prize is awarded to the team that has demonstrated the most progress. CAMTech organizes innovation workshops and entrepreneur boot camps that are run by professionals from industry, investment, clinical, and technical fields—event goals may include product engineering and realization, clinical validation and testing, business strategy, IP and regulatory strategy, financing and investment, and commercialization strategies. CAMTech is also continuously working to assess the value of hackathons. One means of accomplishing this is through follow-up surveys whose results will be presented in future publications.

8.3.3 Case Studies

Co-creation Labs

The staffing and initiatives of the CAMTech Co-creation Lab in Mbarara illustrate the influence that health hackathons in LMICs have had on the local innovator community. The Co-creation Lab manager is a computer engineer and alumnus of the Mbarara University of Science and Technology (MUST) who is a former participant in CAMTech health hackathons. Many of the staff at the Co-creation Lab have been introduced to the principles of co-creation by participating in workshops and hackathons held by CAMTech locally and abroad. This has increased their confidence in their capacity to create novel medical technologies applicable to their setting and potentially appropriate for reverse innovation. Examples include a locally made hand sanitizer to aid with infection control called *Sanidrop*, a modular continuous airway pressure device (*mCPAP*) currently under development in the Co-creation Lab, as well as a digital infusion monitor and control device, a wireless physiological monitor, and an automatic surgical suction pump controller ("CAMTech Uganda,"). The Co-creation Lab has recently launched the MUST Innovation Cafes and the Students Innovation Internship Program, both intended to foster interdisciplinary creativity and innovative thinking in healthcare by MUST students ("CAMTech Uganda,").

Startup Ventures

O2-Matic, one of the winning teams of the 2015 Jugaadathon held in Bangalore, comprised engineers, clinicians, entrepreneurs, and industrial designers. This team developed the concept for a low-cost method for oxygen production to address unpredictable availability of medical gases in limited resource settings. *O2-Matic* won a Post-Hackathon Award, sponsored by the Federation of Indian Chambers of Commerce & Industry (FICCI) and Terumo, to continue working on their project. They used this funding to iterate on their proof-of-concept prototype and are currently filing for intellectual property and pitching their idea to potential investors with the hopes of starting a company in the future.

Startup companies are not just an aspiration for health hackathon winners; several successful ventures in fact got their inception at hackathons. The *Augmented Infant Resuscitator* (*AIR*) was a project started at a health hackathon in Boston jointly organized between CAMTech and MIT Hacking Medicine by a team including an engineer from Canada and physicians from the US and Uganda. The group has raised hundreds of thousands of dollars in grant money and has started to conduct tests at clinics in the field. *AIR*, though designed for a local problem in Uganda, also has a large potential market in the US and Europe. Multiple successful startups coming out of MIT Hacking Medicine's health hackathons can be found in resource-rich settings; they include *Smart Scheduling*, *Podimetrics*, *PillPack*, *Cake,* and *Twiage*.

8.4 Conclusions and Future Directions

Involving all stakeholders in a local context can catalyze innovation for and by the bottom of the pyramid. This provides an opportunity to rethink how we approach global health, shifting the focus to co-creating accessible and affordable solutions, harnessing the potential of emerging markets, and seeking applications of innovative ideas to better health outcomes elsewhere (Mauser et al. 2013; Witty 2011).

Although one of the most exciting outcomes of a hackathon could be the start of a successful company, we acknowledge that this is a rare occurrence. It is not our intent to imply that the primary goal of a hackathon is to start a company. Rather, at the core of these events is exposure to innovation and the establishment of new networks. We aim to inspire local stakeholders to attempt solving problems that intimately affect their lives and to form a network of like-minded individuals.

Hackathons can be criticized for generating short-lived project ideation but no sustainable solutions. We are working on strategies to maintain innovator stamina as teams go through all stages of the solution development process (De Passe et al. 2014; Gardner et al. 2007). Resources that allow teams to work on their projects after the hackathon are key. CAMTech is providing this support to hackathon-spurred global health innovations via awards for teams that make the most progress after the hackathon, Innovation Awards, and the Online Innovation Platform. While Research and Development for LMICs has frequently lacked significant investment by large multinational corporations, the hackathon movement's promise of open-source innovation has encouraged forward-thinking life sciences companies to engage in health hackathons around the globe (e.g., the GE Research and Development Center and the Biocon Foundation in Bangalore, India) (Morel et al. 2005a, b; Witty 2011).

Health hackathons can serve as a launching point that engages the community and demonstrates the existence of passion and ability to solve local problems in culturally appropriate and sustainable ways with promise to improve health outcomes globally.

Acknowledgements The authors would like to thank MIT Hacking Medicine, MGH Center for Global Health, and CAMTech for their guidance and support of this work. They would also like to thank the MIT Institute for Medical Engineering and Science for fellowships to join the CAMTech team in health hackathons in India.

References

Bailey, E. (2014). Hackathons aren't just for coders. We can use them to save lives. *Wired*.

Caputo, I. (2014). Hacking to improve healthcare. *WGBH News*.

Chilukuri, S., Gordon, M., Musso, C., & Ramaswamy, S. (2010). *Design to value in medical devices*. McKinsey & Company.

DePasse, J. W., Carroll, R., Ippolito, A., Yost, A., Chu, Z., & Olson, K. R. (2014). Less noise, more hacking: how to deploy principles from MIT's hacking medicine to accelerate health care. *International Journal of Technology Assessment in Health Care, 30*(03), 260–264.

Gardner, C. A., Acharya, T., & Yach, D. (2007). Technological and social innovation: A unifying new paradigm for global health. *Health Affairs (Millwood), 26*(4), 1052–1061.

IDEO. (2015). *The field guide to human-centered design*.

Lee, S. M., Olson, D. L., & Trimi, S. (2012). Co-innovation: Convergenomics, collaboration, and co-creation for organizational values. *Management Decision, 50*(5), 817–831.

Mauser, W., Klepper, G., Rice, M., Schmalzbauer, B. S., Hackmann, H., Leemans, R., et al. (2013). Transdisciplinary global change research: The co-creation of knowledge for sustainability. *Current Opinion in Environmental Sustainability, 5*(3), 420–431.

Morel, C., Broun, D., Dangi, A., Elias, C., Gardner, C., et al. (2005a). Health innovation in developing countries to address diseases of the poor. *Innovation Strategy Today, 1*(1), 1–15.

Morel, C. M., Acharya, T., Broun, D., Dangi, A., Elias, C., Ganguly, N., et al. (2005b). Health innovation networks to help developing countries address neglected diseases. *Science, 309*(5733), 401–404.

Prahalad, C. K., & Ramaswamy, V. (2004). Co-creation experiences: The next practice in value creation. *Journal of interactive marketing, 18*(3), 5–14.

Sachs, J. (2003). The global innovation divide. *Innovation Policy and the Economy, 3*, 131–141 (MIT Press).

Witty, A. (2011). New strategies for innovation in global health: A pharmaceutical industry perspective. *Health Affairs (Millwood), 30*(1), 118–126.

Chapter 9
Developing a Low-Cost, Ultraportable, Modular Device Platform to Improve Access to Safe Surgery

Debbie L. Teodorescu, Dennis Nagle, Sashidhar Jonnalagedda, Sally Miller, Robert Smalley and David R. King

9.1 Introduction

9.1.1 Surgical Care as Part of the Global Health Armamentarium

Over 30% of the global disease burden requires surgical therapy, which could prevent over 18 million deaths and save USD $200 billion annually. The conditions amenable to surgical therapy range broadly, from traumatic to obstetrical to infectious to oncological and beyond. Yet, in low–middle-income countries (LMICs), an estimated two billion people have effectively no access to surgical care, and another

D. L. Teodorescu (✉) · D. Nagle
MIT D-Lab, Cambridge, USA
e-mail: DLTeodor@surgibox.org; debbiepl@mit.edu

D. Nagle
e-mail: dennisnagle8@gmail.com

S. Jonnalagedda
Program Essential Tech Cooperation and Development Center EPFL, Lausanne, Switzerland
e-mail: sashidhar.jonnalagedda@alumni.epfl.ch

S. Miller
MIT Department of Mechanical Engineering, Cambridge, USA
e-mail: millersa@mit.edu

R. Smalley · D. R. King
Harvard Medical School, Boston, USA
e-mail: rob@surgibox.org

D. R. King
e-mail: dking3@mgh.harvard.edu

D. R. King
Massachusetts General Hospital Department of Surgery, Boston, USA

S. Hostettler et al. (eds.), *Technologies for Development*,
https://doi.org/10.1007/978-3-319-91068-0_9

two to three billion have access only to surgeries performed in unsterile settings such as general-use buildings or even outdoors ("Global Surgery 2030", 2015; Disease Control Priorities Project 2008). In addition to this chronic deficiency in surgical access, field surgical zones in disaster-affected areas are often exposed to frank particulate and insect contamination.

9.1.2 Patient Safety in Surgery: Infrastructural Challenges to Sterility

In LMICs, surgical patients develop disproportionate rates of surgical site infections (SSIs), particularly the deep infections characteristic of intraoperative contamination. Meta-analyses (Allegranzi et al. 2011) have found that 0.4–30.9 per 100 surgical patients in LMICs develop SSIs. In particular, even in clean and clean-contaminated wounds, which had not previously been contaminated by traumatic skin breaks, uncontrolled gut flora spillage, etc., the median cumulative incidences were still, respectively, 7.6% (range 1.3–79.0%) and 13.7% (1.5–81.0%), all several times higher than in higher income countries (Ortega et al. 2011). Most alarmingly, these figures represent early postoperative infections of deep visceral spaces and organs, not superficial tissues, a finding underscored by Nejad et al. (2011) meta-analysis that showed 6.8–46.5% incidence of deep infections in postoperative patients, and 10.4–20.5% of infections in organ spaces. Bjorklund et al. (2005) analysis showed a particularly unfortunate interaction between immunosuppression—all too common in the developing world due to poor nutrition, untreated illness, and HIV—and unsterile surgical conditions in producing very high rates of severe infection following c-sections. These infections translate into longer stays at already-overcrowded hospitals: eight additional days on average in Tanzanian and Ethiopian studies, 10 days in a Burkina Faso study comparing surgical patients with and without SSIs (Eriksen et al. 2003; Taye 2005; Sanou et al. 1999). In nascent healthcare systems with limited infrastructures, SSIs that effectively double or triple patient stay lengths fetter institutions' ability to cope and reduce the volume of new patients that could be accommodated. Taye (2005) noted that SSIs were associated with 2.8-fold increased mortality (10.8% vs. 3.9%).

Numerous factors impact surgical site infection rates. These have been most authoritatively summarized by the Lancet Commission on Global Surgery (2015) and range from preoperative antibiotic administration to drape selection to handwashing and beyond. A particularly pernicious and challenging one to address has been that of the contaminated environment. Whyte et al. (1982) and Edmiston et al. (2005) have described the general link between airborne contamination and SSIs, with an estimated 30–98% of wound bacteria attributable to airborne contaminants, depending on the ventilation system in an operating room. In higher income countries, invasive procedures are typically performed by scrub-attired personnel striving to reduce contamination in operating rooms with meticulously filtered air. In LMICs, such facilities and infrastructure are frequently unavailable. Procedures instead often occur wherever dedicated space could be found, whether general-use rooms, outdoors, or other suboptimal settings. Pathogen-carrying insects, dust particles, provider skin squames, and numerous other dangers frequently breach the sterile field. Even in state-of-the-art operating rooms, relatively modest breaches due to events such as door openings have been associated with increased SSI rates. Indeed, decreasing the number of times doors was opened decreased SSI rates by 36% in one study and 51% in another (Van der Slegt et al. 2013; Crolla et al. 2012). The absence of any door at all, or of effective surgical suite ventilation, in the LMIC operating space is therefore quite a concern.

The need to provide safe surgical care outside of traditional surgical facilities is certainly not a new problem. However, solutions to the challenge have typically started from the assumption that the core problem is to provide a sterile operating room outside of a standard facility. This mindset informs solutions such as surgical tents, operating rooms mounted on trailers or trucks, semi-portable laminar airflow systems, and most other solutions to date. These devices unfortunately tend to share several significant limitations in practice. They are challenging to transport to remote or disaster-affected areas, requiring both time and logistical capability. Once at the desired site, they require significant setup time. For example, surgical tents can take a full team of technicians working around the clock for 72 h to fully set up. Several of these systems have at least one external dependency, particularly availability of electricity or requirement of flat terrain. They require significant resources not only for sunk cost but also for marginal cost of each procedure. Personnel, a particularly scarce resource, is also required to set up and maintain these complex systems. These systems are not always robust to the high levels of external contamination, with sand particulate ingress into the tents a particularly notorious phenomenon in the field (e.g., as described by Stevenson and Cather 2008). Finally, any contamination in these systems, including that generated by providers through squame shedding, can still contaminate the surgical site.

9.1.3 Provider Safety in Surgery: Protecting Surgical Teams

Patients are not the only ones who can get infected during surgeries. Some 85,000 medical providers worldwide are infected every single year by patient bodily fluids, with the vast majority of surgeons and obstetrician/gynecologists having experienced at least one exposure in the past year (Butsashvili et al. 2012). Despite the lower volume of invasive procedures occurring in austere settings, 90% of providers infected were working in such settings (World Health Organization 2011). Such chronic risks were thrust into sharp relief during the Ebola epidemic, when, for instance, Sierra Leone's surgeons encountered 100-fold infection rate increases compared with the general population, resulting in the death of 25% of the surgeons in the main teaching hospital of the capital (Yasmin and Sathya 2015; Bundu et al. 2016). Unfortunately, personal protective equipment (PPE) is costly and cumbersome to wear during surgeries, leading to both poor availability and poor provider adherence. Thus, surgical teams are vulnerable to infections from patient bodily fluids.

9.1.4 SurgiBox: Solution Concept for the Double Challenge in Safe Surgery

The SurgiBox platform moves away from the assumption that the surgical space of interest is the operating room. Fundamentally, the space that matters is the incision and immediate surgical field over the patient. This recognition literally shrinks the challenge down from over thousands of cubic feet of space to well under 10 cubic feet to be kept sterile. The nature of protection actually changes, as contamination can come from patients themselves, providers, and the external environment. Creating a physical barrier from contamination completely from the latter two, and significantly from the patients themselves, theoretically permits more robust protection than would the classic, costly combination of sterile room, sterile scrub suit, and sterile drapes. A system that can effectively maintain a sterile field in this limited space, as presented here, provides a low-cost, ultraportable platform for regulating intraoperative conditions at the surgical site, making safe surgery more accessible. At the same time, isolating the surgical field blocks potentially infectious particles and fluids from reaching providers at all. This is a more efficient system than capturing with individual PPE after they have already left the surgical field.

This paper is organized as follows. Section 9.2 describes SurgiBox's iterative prototyping and evaluation methods. Section 9.3 presents the design and our results to date. Section 9.4 discusses ongoing as well as future efforts in device development and deployment. Section 9.5 discusses SurgiBox's broader implications.

9.2 Methods

9.2.1 Patient- and Stakeholder-Centered Development

The objective of SurgiBox overall is to provide a low-cost, ultraportable system to maintain sterile conditions during invasive procedures, even when performed in contaminated settings. However, as with any medical device, particularly one intended for developing settings, other complex requirements are critical to stakeholder acceptance (Caldwell et al. 2011). A key risk was the rejection of the device by the end user despite technical success, so extensive measures were taken to mitigate this. Extensive preliminary interviews as well as ongoing stakeholder interviews were conducted, and input was received from physicians who work in the developing world, surgical researchers, biomedical and device engineers, global health and development researchers as well as other workers, members of the medical device industry,

Table 9.1 Stakeholder-generated device specifications

Ultraportable	Entire system should fit into a backpack or duffel bag
Quick setup	Setup time should not interfere with surgical prep
Ergonomic	Should fit well into existing surgical workflow
One size fits all	Can be used for all sizes of patients by all types of users
Low cost	Cost should not exceed the cost of surgical drapes
Self-contained	Battery-powered
Sterile	Meets or exceeds operating room standards
Good visibility	User's view of surgery must be unobstructed
Protective	Prevents bodily fluid splashes and aerosols from reaching users

and innovation strategists. These discussions generated the objectives and specifications as shown in Table 9.1. We used existing data from anthropometric tables (with the aim to accommodate the 5th through 95th percentiles of providers and of patients), surgical ergonomics research, and operating room design guidelines to populate design specifications for the prototypes.

Throughout the design process, the prototype has been split into modules to improve team efficiency. The overall design concept was split into enclosure design, ports design, and environmental control system design.

9.2.2 *Proof of Concept Testing*

In addition to evaluating ergonomic and workflow acceptability, we focused on whether the system actually provides a level of sterility consistently equivalent to or exceeding that available in state-of-the-art operating rooms.

We actually proved this for two separate setups of SurgiBox, both set up in a mixed-use machine shop at MIT D-Lab. In an earlier iteration, as reported in Teodorescu et al. (J Med Dev 2016), the prototype utilized a rigid external frame and therefore started with a full internal volume of contaminated air. The environmental system was based on an off-the-shelf powered air purifying respirator system calculated to supply 110 air changes per hour from a simple hole-in-side inlet. Measurements were then taken at the xiphoid as the approximate center point of a large surgery combining laparotomy and thoracotomy as may occur to address trauma or hemorrhage, as well as at the flanks to assess particle pooling. These were repeated with armports and material ports open. In the more recent iteration, as detailed in Teodorescu et al. (in press, IEEE Xplore), the enclosure is inflated from flat packaging, but intentionally not sterile as it would be in real life, to mimic contamination that could occur during introduction of the instrument tray during setup. Air was supplied at 66 air changes per hour by a HEPA-motor-power setup that we built ourselves. The system then used a special manifold setup to distribute airflow in laminar fashion through the enclosure. Material ports were kept closed; armports were kept in neutral position. Particle counts were then benchmarked against Wagner et al. (2014) data correlating particle counts and colony-forming units in operating rooms.

9.3 Results

9.3.1 Device Design

SurgiBox is an ultraportable, modular system to provide sterile intraoperative environments over surgical sites. This product went through extensive evolution. Initially, the design comprised a reusable box-like system with hinged clear panels of polycarbonate that could be collapsed into a flat package. Based on stakeholder feedback regarding sterilization capabilities, bulk, ergonomics, and modularity potential, the design was iterated upon and became a disposable, patient-contacting plastic enclosure with a minimal frame, and reusable environmental controls. The arm ports were redesigned several times to ensure that the system was ergonomic for the users and minimized the chances of contamination. By contrast, whereas we had originally planned for material ports to be hermetic airlock-inspired systems, feedback on impact to workflow prompted redesign to quick-opening ports with a single layer of sealing, and this design was shown to not compromise sterility through environmental testing. Throughout the design process, each iteration was tested with local surgeons, ensuring that each new design improved upon the previous.

The final design comprises a low-density polyethylene enclosure with a high-visibility vinyl window. The enclosure adheres to the patient's surgical site with an adhesive iodine-infused antimicrobial drape and is inflated with HEPA-filtered air. There is an optional minimal frame that can be used with the enclosure, but the enclosure can also be used alone and the positive pressure within provides its structure. In the case that a port needs to be opened to move materials in or out, the airflow can simply be increased to accommodate the extra outflow. The inflow system is designed such that the filtered air enters the enclosure directly over the surgical site, in the same way that air is introduced to state-of-the-art hospital operating rooms.

There are four sets of arm ports to accommodate up to four users at any given time. The arm ports were designed so the users can perform the same motions that they would use in a standard operation, and can pass tools back and forth easily. There are also four material ports so materials (such as tools or even an infant in the case of a cesarean section) can quickly be passed in or out of the surgical field.

Based on serial testing of workflow, it was found that the time of patient being positioned on the operating table to time of first incision was consistently less than 85 s for users naïve to the system. Per qualitative report from the medical members of the team, this timing compares favorably to that of existing patient and provider preparation. We note that system does not preclude users from donning additional

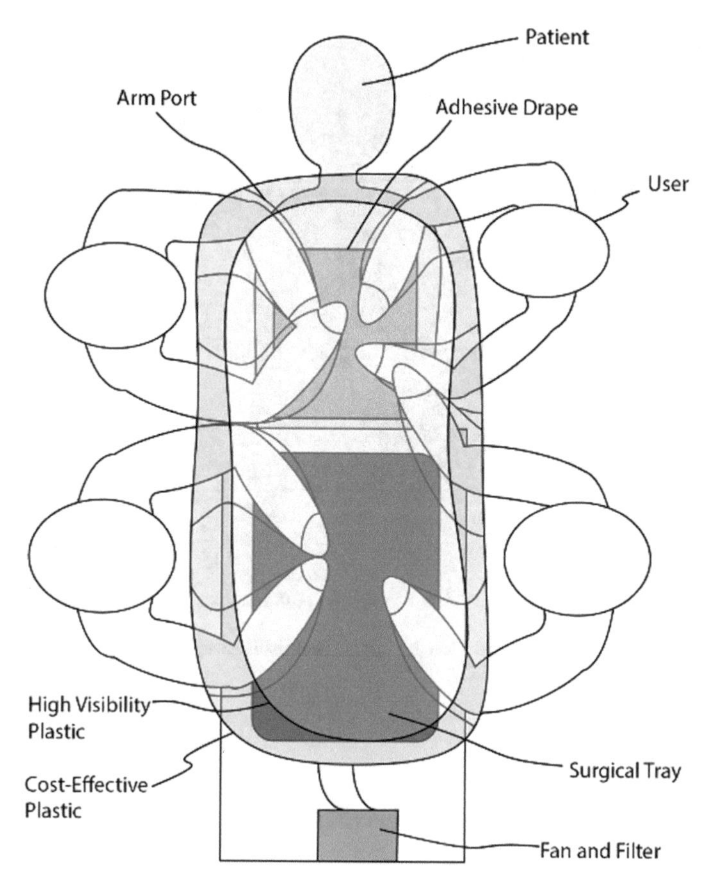

Fig. 9.1 Schematic of surgeons and nurses performing a surgical procedure with SurgiBox

personal protective equipment or further draping the patient but reduces the burden of both (Fig. 9.1).

A rechargeable battery powers a microcontroller and fan so the system can be used off-grid. The airflow is high initially for rapid system inflation but is subsequently turned down to a maintenance flow to maintain the pressure inside the enclosure (Fig. 9.2).

9.3.2 *Particle Testing*

The rigid frame system consistently achieved 0 particle count at all test sites within 90 s, or after 2.75 air changes. The minimal frame system consistently achieved target thresholds even before completion of insufflation. Importantly, both systems maintained acceptable thresholds with open ports.

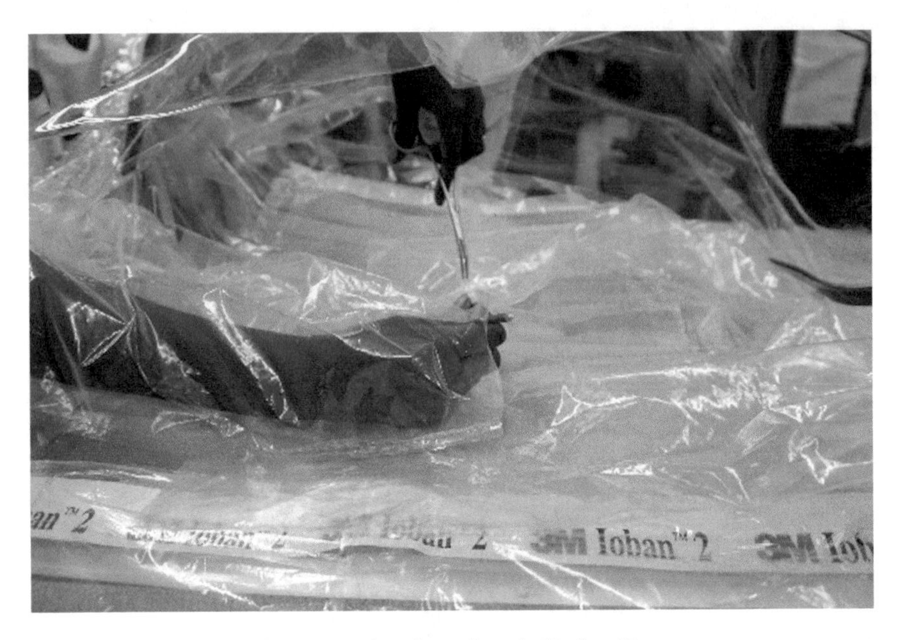

Fig. 9.2 Range of motion demonstrated as shown from inside SurgiBox

9.4 Discussion

9.4.1 Ongoing and Future Research

Our efforts to further optimize SurgiBox with our existing iterative prototyping workflow are ongoing. At the time of submission, we are completing two major prototyping tasks in parallel. First, we are preparing advanced human factors testing with full-length simulated procedures. Second, we are optimizing for manufacturability. To date, the latter efforts have yielded over 12-fold decrease in system cost. The cost-determining portions are fully reusable between patients for an estimated 10,000 cases depending on airflow requirement, based on the life expectancies of the battery and the filter cartridge, both of which are replaceable to permit continued use of the system.

The patient-contact components are ideally disposable because in many cases setting up reliable reprocessing can in fact be more logistically challenging than stockpiling disposable components, especially ones of minimal bulk. In any case, once the design has been optimized, it will be important to continue work with medical logisticians and hospital administrators to estimate the setup and recurring costs that constitute the total cost of ownership needed to meet each institution's needs, based on procedures and personnel.

In addition, the current experimental setup has two main limitations. First, although particle testing is a well-validated way to obtain dynamic measurements of the system's barrier function, we also plan to correlate with settle plate testing to assess colony-forming unit counts. The second limitation is that the tests are conducted in static conditions. Ongoing work is using standardized rigs to simulate use conditions.

While we do strive to simulate field conditions in the lab and to assess ease of use of the prototypes by stakeholders with experience in LMIC settings, we are planning to work with LMIC community partners abroad to supply SurgiBox kits to the field. By performing ex vivo setups in truly field conditions, we can collect further data and feedback to identify any additional issues requiring optimization as well as impacts on workflow. These will most likely occur in India and Uganda. By supplying SurgiBox kits with a variety of frame and port components, we hope to encourage community partners to tinker with the prototypes to suit their needs.

While we have striven at every step to make SurgiBox as user-friendly as possible, we recognize that there will likely be resistance if it were presented on the basis of in vitro sterility data only. After all, there are many determinants of safe surgery. Therefore, even though SurgiBox would qualify as a CE Mark IIA device that does not require efficacy studies, we still plan to voluntarily conduct trial surgeries on animals in simulated field conditions to assess impact on clinical outcomes such as wound contamination rates and SSI rates.

9.4.2 Road to the Market

The key to deploy SurgiBox worldwide is to understand the existing market, the needs, and how the device can fill the void. Our major considerations fall into three main categories: market segmentation, production, and distribution.

9.4.2.1 Market Segmentation

The first step is to segment the market into different categories. There are two main segments that SurgiBox strives to target: on one hand, it can be used to reinforce protections available in existing medical infrastructures. On the other hand, it can be used as part of an ultraportable kit that gives access to surgery in places where there is no medical care, such as war zones, natural disaster areas, and even remote villages only accessible by foot.

Through different interviews with surgeons from LMICs, we discovered that many existing infrastructures such as district hospitals are austere: due to their very high cost, ventilation systems are often absent from operating rooms, which also often have doors open throughout surgeries, sometimes even directly showing to the main road. All these highly increase the probability for the patients to develop SSIs. SurgiBox is therefore expected to reduce healthcare costs by permitting surgical procedures to be performed in less-expensive procedure rooms, and reduce SSIs if used in tandem with existing facilities. In tertiary care centers, it can conceivably offer improved outcomes in lengthy, complex procedures by providing more intensive control of the intraoperative environment. LMICs will be interested in this device because of its affordable price for creating a sterile environment for surgery that meets or exceeds US and European standards, leading to a general improvement in surgical outcomes.

Within the second market segment, this device addresses a distinctive problem, the unsafe and unsterile ad hoc operative location. Some stakeholders from NGOs shared that they sometimes had to operate in open air. We expect the early adopters to be surgeons from higher income countries working in disaster zones and in low resource settings. While in disaster zones our device is a direct fit as there is no infrastructure at all, in low resource settings it addresses the main concern of their current inability to provide the standard of hygiene and infection mitigation to which they are accustomed. In addition, it also enhances provider safety by offering them a reduction of exposure to the patient's bodily fluids and aerosols. During the Ebola outbreak, physicians and organizations promulgated and implemented standards for appropriate drapes and protective equipment for operating on possibly infected patients. SurgiBox is in line with such protective efforts.

9.4.2.2 Production

To support advanced development of SurgiBox and pilot deployment, our collaboration is able to leverage our position at the intersection of academia and nongovernmental organization to seek both competitive grant funding and mutually beneficial partnerships with other nongovernmental or governmental organizations. As a social venture, we further benefit from the robust small business and social entrepreneurship resource infrastructure available in the United States in general and in the Boston area in particular, including funding initiatives.

SurgiBox's design and prototyping process have emphasized minimizing not only the cost of raw materials but also of manufacturing and packaging. By finding alternatives to complex, high-variance, high-cost components, we expect that the device should be able to be manufactured to consistently high quality to comply with United States Federal Drug Administration General Controls and similar regulations. Indeed, it is conceivable that we can eventually engage with regional or local manufacturing entities during scale-up stage, recognizing the importance of maintaining quality controls compatible with a medical device. With the "cost of goods sold" for benchtop prototypes stabilizing, we are now conducting this analysis in the real world setting.

9.4.2.3 Distribution

SurgiBox itself will be distributed as ultraportable, fully self-contained, ready-to-use kits suitable for hand luggage, backpacks, drones, and other limited spaces. These kits will contain the reusable component, one or more of the patient-contacting components all individually wrapped, and batteries: all the items needed for full use of the system. Users can therefore continue to utilize their preferred skin disinfectant, lighting source, instrument trays, gloves on top of the universal-sized thinner gloves in place, and other things to best preserve workflow.

At this stage, we are working on finalizing strategic partnerships critical to pilot deployment and eventually distribution success in the future. Supply and demand bottleneck analyses of the expected uptake challenges along the value chain are ongoing, as highlighted above. In the market segment we are first targeting, procurement is primarily by each mission-sponsoring entity—most commonly militaries, surgical relief organizations, hospitals, and device companies—or by individual providers. Upstream, many of the former have relationships with procurement superstructures such as the World Health Organization, which in 2015 reported allocating the plurality ($333 million) of its procurement budget to strategic category products, which cover most key surgical devices, tools, and kits. For the second market segment, we plan to contact Ministries of Health and Defense in LMICs. Certainly, engaging with all of these diverse stakeholders is critical to success.

9.5 Conclusion

Taken together, the growing interest in surgery as an inalienable part of global health, as well as the ethical as well as practical need to provide this surgical care in a safe manner, provides a rich opportunity for innovative solutions to the complex challenges entailed.

In this paper, we described one such innovation in the form of a co-designed, ultraportable sterile field platform. By shifting the site of regulation from the operating theater to the incision itself, we introduced a novel paradigm more amenable to flexible, cost-effective solutions.

To reduce this paradigm to practice, we closely engaged user–stakeholders by starting with a systematic needs analysis, then using feedback to drive the evolution and refinement of SurgiBox. We presented the device design and results from benchtop testing that showed how SurgiBox can rapidly create a particle-free environment. Ultimately, deploying SurgiBox to LMICs and beyond requires continued close stakeholder engagement in the form of robust relationships along the production and supply chains.

Acknowledgements We thank D-Lab's Amy Smith and Victor Grau Serrat (now of Color Inc) for consistent support over the years. Stephen Odom and Dana Stearns have provided key design input and implementation insights. We are further grateful to Marissa Cardwell, Jim Doughty, Fabiola Hernandez, and Kerry McCoy of MIT's Environmental Health Services for technical assistance and device design feedback. Kristian Olson offered invaluable advice on implementation and deployment strategy. Jack Whipple has provided machine shop assistance. Christopher Murray, Thomas Shin, and Robert Smalley have provided design and developmental input. This work was supported by the Harvard Medical School Scholars in Medicine Office.

References

Book Chapter

American Society of Heating, Refrigeration and Air-Conditioning Engineers. (2011). Health Care facilities (I-P). In *ASHRAE 2011 handbook—HVAC application*. Atlanta: ASHRAE.

Journal Article

Allegranzi, B., Bagheri Nejad, S., Combescure, C., Graafmans, W., Attar, H., Donaldson, L., et al. (2011). Burden of endemic health-care-associated infection in developing countries: A systematic review and meta-analysis. *Lancet, 377*(9761), 228–241.

Amenu, D., Belachew, T., & Araya, F. (2011). Surgical site infection rate and risk factors among obstetric cases of Jimma University Specialized Hospital, Southwest Ethiopia. *Ethiopian Journal of Health Sciences, 21*(2), 91–100.

Bickler, S. N. et al. (2015). *Essential surgery: Disease control priorities* (3rd ed., Vol. 1). Washington: The World Bank.

Bjorklund, K., Mutyaba, T., Nabunya, E., & Mirembe, F. (2005). Incidence of postcesarean infections in relation to HIV status in a setting with limited resources. *Acta Obstetricia et Gynecologica Scandinavica, 84,* 967–971.

Bundu, I., Patel, A., Mansaray, A., Kamara, T. B., & Hunt, L. M. (2016). Surgery in the time of Ebola. *Journal of the Royal Army Medical Corps, 162*(3), 212–216.

Butsashvili, M., et al. (2012). Occupational exposure to body fluids among health care workers in Georgia. *Occupational Medicine, 62*(8), 620–626.

Caldwell, A., Young, A., Gomez-Marquez, J., & Olson, K. R. (2011). Global health technology 2.0. *IEEE Pulse, 11,* 63–67.

Crolla, R. M., van der Laan, L., Veen, E. J., Hendriks, Y., van Schendel, C., & Kluytmans, J. (2012). *PLoS ONE, 7*(9), e44599.

Edmiston, C. E., Seabrook, G. R., Cambria, R. A., et al. (2005). Molecular epidemiology of microbial contamination in the operating room environment: Is there a risk for infection. *Surgery, 138*(4), 573–582.

Eriksen, H. M., Chugulu, S., Kondo, S., & Lingaas, E. (2003). Surgical-site infections at Kilimanjaro Christian Medical Center. *Journal of Hospital Infection, 55*(1), 14–20.

Klingler, G. A. (1972). Digital computer analysis of particle size distribution in dusts and powders. Resource document. *National Technical Information Service*. http://www.dtic.mil/cgi-bin/GetTRDoc?Location=U2&doc=GetTRDoc.pdf&AD=AD0752209. Accessed 20 October 2015.

Meara, J. G., et al. (2015). Global surgery 2030: Evidence and solutions for achieving health, welfare, and economic development. *Lancet, 386*(9993), 569–624.

Nejad, S. B., Allegranzi, B., Syed, S. B., Ellis, B., & Pittet, D. (2011). Health-care-associated infection in Africa: A systematic review. *Bulletin of the World Health Organization, 89,* 757–765.

Ng-Kamstra, J. S., et al. (2016). Global surgery 2030: A roadmap for high income country actors. *BMJ Global Health, 1*(1), e000011.

Online document. (2008). Promoting essential surgery in low-income countries: A hidden, cost-effective treasure. *Disease Control Priorities Project*. http://www.dcp2.org/file/158/dcpp-surgery.pdf. Accessed February 25, 2012.

Online document. (2015). Global surgery 2030: Evidence and solutions for achieving health, welfare, and economic development. *The Lancet Commission on Global Surgery*. http://www.globalsurgery.info/wp-content/uploads/2015/01/Overview_GS2030.pdf. Accessed May 12, 2015.

Ortega, G., Rhee, D. S., Papandria, D. J., Yang, J., Ibrahim, A. M., Shore, A. D., et al. (2011). An evaluation of surgical site infections by wound classification system using the ACS-NSQIP. *Journal of Surgical Research, 174*(1), 33–38.
Sanou, J., Traore, S. S., Lankoande, J., Quedraogo, R. M., & Sanou, A. (1999). Survey of nosocomial infection prevalence in the surgery department of the Central National Hospital of Ouagadougou. *Dakar Medical (abstract only), 44*(1), 105–108.
Secretariat of the Safe Injection Global Network, World Health Organization. Aide-memoire for a strategy to protect health workers from infection with bloodborne viruses. (2011). World Health Org. WHO/BCT/03.11.
Stevenson, K., & Cather, C. (2008). Pursuing cleanliness in a field surgical environment. *AORN Journal, 87*(2), 306–309.
Taye, M. (2005). Wound infection in Tikur Anbessa hospital, surgical department. (2005). *Ethiopian Medical Journal (abstract), 43*(3), 167–174.
Teodorescu, D. L., Miller, S. A., Jonnalagedda, S. (2007). SurgiBox: An ultraportable system to improve surgical safety for patients and providers in austere settings. IEEE Xplore GHTC 2017 (accepted, pending publication).
Teodorescu, D. L., Nagle, D., Hickman, M, King D. R. (2016) An ultraportable device platform for aseptic surgery in field settings. *Journal of Medical Devices, 10*(2), 020924 (May 12, 2016).
Van der Slegt, J., Van der Laan, L., Veen, E. J., Hendriks, Y., Romme, J., & Kluytmans, J. (2013). *PLoS ONE, 8*(8), e71566.
Wagner, J. A., Schreiber, K. J., & Cohen, R. (2014). Improving operating room contamination control. *ASHRAE., 56*(2), 1–10.
Whyte, W., Hodgson, R., & Tinkler, J. (1982). The importance of airborne bacterial contamination of wounds. *Journal of Hospital Infection, 3,* 123–135.
Yasmin, S. & Sathya, C. (2015). Ebola epidemic takes a toll on Sierra Leone's surgeons. *Scientific American*. 2015.

Part IV
Renewable Energies

Chapter 10
Rural Electrification and Livelihood Generation for Women Enterprises in Rural India: Experience of Implementing Two-Stage Biomass Gasifiers

Sunil Dhingra, Barkha Tanvir, Ulrik Birk Henriksen, Pierre Jaboyedoff, Shirish Sinha and Daniel Ziegerer

10.1 Introduction

As part of the project, a two-stage gasifier of 20 kWe capacity, with simple cleaning and cooling system, has been developed, tested and localized to Indian conditions. The experience so far has demonstrated high-quality gas (tar < 25 mg/m^3), low specific fuel consumption, and no wastewater generation in cleaning and cooling. Table 10.1 shows the efficiency of the two-stage gasifier system developed in cooperation between The Energy and Resource Institute (TERI) and Technical University of Denmark, and Fig. 10.1 shows the process flow diagram of the system.

S. Dhingra (✉) · B. Tanvir
The Energy and Resources Institute (TERI), New Delhi, India
e-mail: dhingras@teri.res.in

B. Tanvir
e-mail: barkhatanvir@gmail.com

U. B. Henriksen
Technical University of Denmark (DTU), Copenhagen, Denmark
e-mail: ubhe@kt.dtu.dk

P. Jaboyedoff
Effin Art, Lausanne, Switzerland
e-mail: pierre.jaboyedoff@effinart.ch

S. Sinha · D. Ziegerer
Swiss Agency for Development and Cooperation (SDC), New Delhi, India
e-mail: shirish.sinha@eda.admin.ch

D. Ziegerer
e-mail: daniel.ziegerer@eda.admin.ch

S. Hostettler et al. (eds.), *Technologies for Development*,
https://doi.org/10.1007/978-3-319-91068-0_10

Table 10.1 Efficiency improvements in two-stage gasifier

Key parameters	Fixed-bed gasifier	Two-stage gasifier	Improvements
Energy efficiency			
Overall efficiency (in terms of biomass consumption to electricity generated)	15%	>22%	About 50% efficiency gain due to reduction in parasitic loads and reduction in fuel consumption
Resource use efficiency			
Input fuel moisture—quality of biomass that can go as an input	up to 15%	up to 30%	Greater flexibility in terms of biomass quality (dry) and use
Environmental benefits			
Tar (raw gas)	500–600 mg/Nm^3	Less than 25 mg/Nm^3	Impurities in the raw gas in form of tar significantly reduced. This improves the overall environment performance of the system as less tar is generated
Gas cooling and cleaning system	Water scrubbing—350 L of fresh water for 50 h of operation	Without water scrubbing	Zero water usage for removing impurities in the gas
Wastewater generation	350 L after 50 h of operation	No wastewater generation	

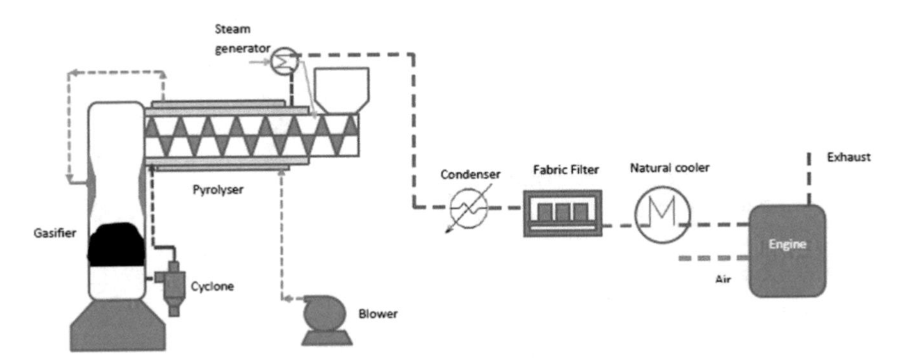

Fig. 10.1 Two-stage gasifier system process flow diagram

10.2 Research Objectives and Involvement of Public Sector

Lessons learnt from past experiences showed the need for a robust business model to support the economics of running of a gasifier with minimal maintenance and suitability for field implementation. Due to low income and high electricity subsidy in rural areas through conventional grid system, it is becoming challenging to support renewable energy-based electrification projects, particularly based on biomass resource on account of high fuel and labour costs. Utilizing the energy from the gasifier for only rural household electrification led to non-payment towards the functioning of the gasifier due to lack of incentive (World Bank 2011).

The project has developed a new approach with focuses on developing business models by combining energy-driven economic activities and meeting household energy requirements in villages in regions of India. With community-based activity getting regular electricity to run successfully, the operation cost of the gasifier towards fuel and operator salary is met and, hence, can also electrify the villages around it. The project model is such that the gasifier will run during daytime providing electricity to the micro-enterprises and in the evening, when lighting needs of households are highest, the gasifier will supply electricity to the households in the nearby villages.

The project has identified around 20 potential sites in the states of Odisha, Jharkhand and Madhya Pradesh, and shortlisted four sites for field implementation of two-stage gasifier through partnerships with public institutions, such as Odisha Renewable Energy Development Agency (OREDA), Rural Electrification Cooperation and Madhya Pradesh State Bamboo Mission (MPSBM) for supporting the investment in implementation covering both utility grid-connected and off-grid villages in the vicinity of these potential sites. The work is also initiated with Odisha Electricity Regulatory Commission (OERC) for obtaining regulatory permission to supply electricity to nearby villages through existing utility grid with operational agreement and for higher electricity tariff based on small capacity biomass gasifier technology to meet the operational cost of these systems and make them economically viable. TERI has signed a Memorandum of Understanding with a local partner organization and owner of the micro-enterprise at each of the sites for support during and after implementation.

The experience of adapting two-stage biomass gasifier for Indian condition and context has so far demonstrated robustness of the technology with minimal maintenance and suitability for field implementation. The future now is to demonstrate a viable business model and regulatory support with local institutional involvement to bring sustainable electricity supply to grid-connected and off-grid villages.

10.3 Research Methodology

The methodology included consultation with public sector agencies, NGOs, district administration and site selection for implementation. The selection of sites in the three states was done with consultations stakeholder, review of secondary sources and determination of criteria, such as economic loads in the cluster, electricity availability in the villages around, presence of local NGO, biomass availability and cost, undertaking primary surveys, and analysis of collected information. Secondary data helped narrow down a few districts on the basis of status of electrification of villages. Each state's un-electrified villages/economic clusters were mapped in consultation with state-level agencies, such as OREDA, National Rural Livelihoods Mission, Jharkhand Renewable Energy Development Agency, MPSBM, MP District Poverty Initiatives Project (MPDPIP), Odisha Tribal Empowerment & Livelihoods Programme, Department of Rural Development, District Industries Centre, etc.

10.4 Selection of Village Clusters

Identification of livelihood clusters in electricity-deficit areas where production is suffering from irregular supply of electricity or is dependent on diesel-based electricity was undertaken in consultation with local organizations, such as NGOs working for the creating/development of that cluster. Site visits to the selected clusters were made for a scoping study and preliminary surveys to learn status of the livelihood activity, number of villagers involved with the activity, number of households in the villages around and other demographic data, current business model of the activity and its finances, institutional setup of the activity such as the role of the NGO or local organization, community-level groups such as self-help groups (SHGs), etc. The scoping study to find these clusters targeted livelihood activities, such as production of nutrition mix for women and children in Anganwadis, cashew processing, milk chilling and packaging, jute processing, tussar silk reeling, lemongrass oil distillation, poultry feed making, bamboo furniture making, etc. The complete list of livelihood clusters for which scoping study was done is given in Annexure 1. These production units are mostly run by women of several SHGs in the rural areas. There is irregular electricity supply in these remote areas, and the quantity and quality of the product suffer due to it. The two-stage biomass gasifier will provide regular and clean supply of electricity at a low cost, which will increase the production and their income, and this increase will improve the standard of living of the women and their families and also give them an opportunity to explore more activities and build training centres. Four such livelihood clusters given in Table 10.2 have been selected, and an extensive feasibility study has been done for implementation.

The detailed project reports of potential feasible projects were prepared to assess electricity and thermal energy requirements for machinery/process used, current status of electrification, alternate options, and challenges faced due to shortage in the livelihood cluster and payments, biomass availability and surplus details, status of electrification in the households, such as proximity of the surrounding villages and number of households for electrification, land ownership and availability for the installation of the gasifier, willingness to pay, investment, and financial calculations. An improved institutional framework has been worked out for technical and social implementation of project, as shown in Fig. 10.2.

Table 10.2 Selected livelihood clusters

S. No.	Activity	Location	Supporting partner
1.	Nutrition mix making	Rayagada district, Odisha	NGO AKSSUS—Adivasi Krushi Swasthya Sikhya Unnayan Samiti
2.	Cashew and agro-processing	Koraput district, Odisha	NGO SPREAD—Society for Promoting Rural Education and Development
3.	Sal leaf plate making	Mayurbhanj district, Odisha	NGO DEEP—Development of Environment and Education Project
4.	Bamboo furniture making	Balaghat district, Madhya Pradesh	Madhya Pradesh State Bamboo Mission

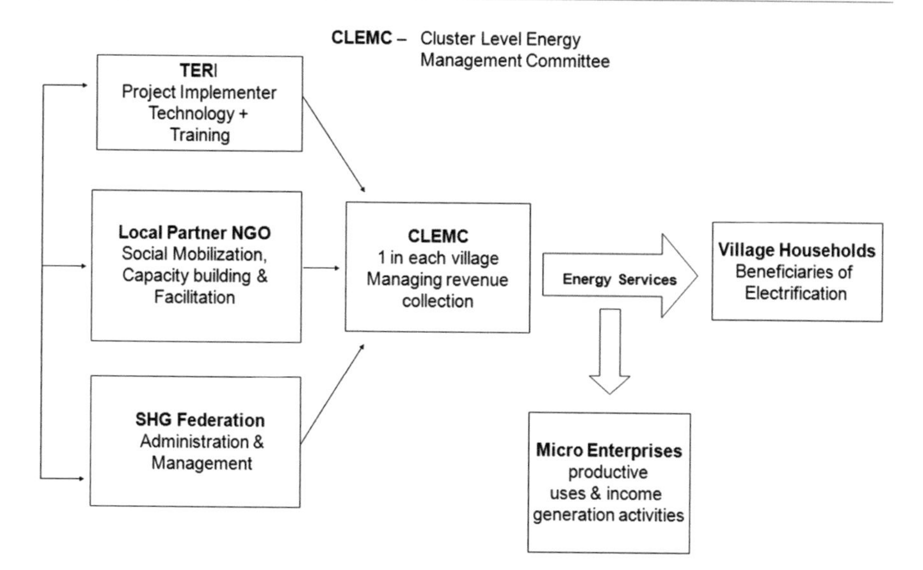

Fig. 10.2 Implementation institutional structure

10.5 Koraput, Odisha

Implementation of the two-stage biomass gasifier in Koraput has started in Odisha. The gasifier will support a new livelihood cluster of processing cashew, pulses and rice to sell directly to the market, which would have not been possible without a regular supply of electricity. As of now, there is no processing happening due to very poor voltage, and the villagers are just drying the shelled cashew for a few days and selling them to middlemen at the weekly market. This does not give them enough income. Processing of the cashew will increase their profit by over 500%. This activity will be owned and run by women SHGs of the villages nearby. A picture

Fig. 10.3 Cashew stored for selling and the local family involved

of cashew stored in gunny bags along with tribal family involved in this activity is given in Fig. 10.3.

With a sustained business model and a self-sufficient power generation system, the community can identify other activities that are possible, which will employ even more women from more villages and increase their incomes in the long run.

The livelihood activity being initiated with the help of the biomass gasifier will be operated and managed by a women's SHG federation called Kasturaba Mahila Maha Sangh. This will give the women employment and an identity outside of their role in taking care of their households. They will be the earning members of their families and have a community support to be independent. These women will be given training on how to use the equipment to process cashew, pulses and rice. They will be taught from start to finish of making the product and packaging it and also on how to manage the finances of the business. The operators selected, women in the SHGs and the villagers have shown eagerness to learn these new skills and also signed a resolution of the same.

10.6 Mayurbhanj, Odisha

Sal leaf plate making in Mayurbhanj in Odisha will start with the implementation of the gasifier as there is no supply of electricity in this village. Once the gasifier is installed, the forest department will support initiation of sal leaf plate making by the SHGs. Pictures of the tribal women involved in sal leaf collection and manual stitching operation are shown in Fig. 10.4. The electricity provided by the gasifier will increase their annual income by 300%.

Fig. 10.4 Tribal women in manual stitching of sal leaf plates activity

10.7 Balaghat, Madhya Pradesh

MPSBM will support TERI in implementing the gasifier for the making of bamboo furniture (as in Fig. 10.5) in rural areas of Balaghat district. Men and women from nearby villages come to work in the factory that makes bamboo furniture. In the process, some bamboo waste is also generated, which can be used as fuel for the gasifier, thus making the factory self-sufficient in the energy requirements.

Under the project, full ownership of the power plant of two-stage biomass gasifier rests with the SHG federation. This body will have the sole responsibility of managing the affairs of the gasifier system for running of the bamboo furniture making factory, which is their sole livelihood activity. With the revenue generated by the sale of electricity and the income from by the sale of their products, they will be in a strong position to pay for the salary of the operator, the biomass supply and maintenance. This will allow them to be completely self-sufficient in their activity.

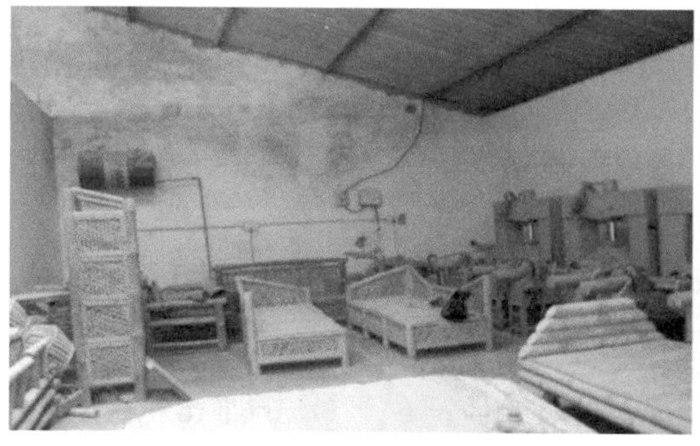

Fig. 10.5 Making of bamboo furniture

10.8 Potential Development Impact

The lack of access to electricity is a major inhibitor to achieving equitable growth and building resilience of poor and vulnerable communities. With a regular supply of electricity in villages, the rural households will have immediate access to basic comforts of lights, fans and charging point, increasing their standard of living. This will gradually help them enhance livelihood opportunities even during nighttime to increase income.

Through the project, TERI and Swiss Agency for Development and Cooperation (SDC) will instal this technology in four livelihood clusters run by women SHGs in villages of Odisha and Madhya Pradesh. The two-stage gasifier of 20–40 kW will support the electrical and thermal needs of the commercial activity and electricity supply in villages around of project site. The livelihood activities being supported and initiated with the help of the biomass gasifier will be operated and managed by a women's SHG federation. This will give the women employment, improve productivity, reduce drudgery, improve health and give an identity outside of the house. They will be the earning members of their families and have community support to be independent.

10.9 Rayagada, Odisha

The nutrition mix (*chhatua*) making is run and owned by the women of an SHG Federation—Maa Dharani Mahila Maha Sangh (MDMMS) in Bissamcuttack block of Rayagada district in Odisha. A Self-Help Group (SHG) is made by women in a village to come together to have regular meetings in which they made savings to open a bank account to have the ability to take loans. These groups were made for women empowerment, a forum where women could feel united to share their problems, find solutions and make ideas for livelihood activities a reality. An SHG Federation is a block level forum where all the SHGs from all the villages in a block come together.

The nutrition mix produced by MDMMS is provided to 75 Anganwadi centres, which then provides this dry take-home ration to pregnant and lactating mothers, infants and children under the Integrated Child Development Scheme of the government. Anganwadi is a government-sponsored child-care and mother-care programme in India. It caters to children in the 0–6 age group. The word 'Anganwadi' means 'courtyard shelter' in Hindi. They were started by the Government of India in 1975 as part of the Integrated Child Development Scheme programme to combat child hunger and malnutrition. The Anganwadi centres provide hot meal to the children and a dry take-home ration. This dry ration is the nutrimix that MDMMS produces for pregnant and lactating mothers, infants and children. The picture showing the women involved in manual cleaning of grains and milling of cleaned grains is shown in Fig. 10.6.

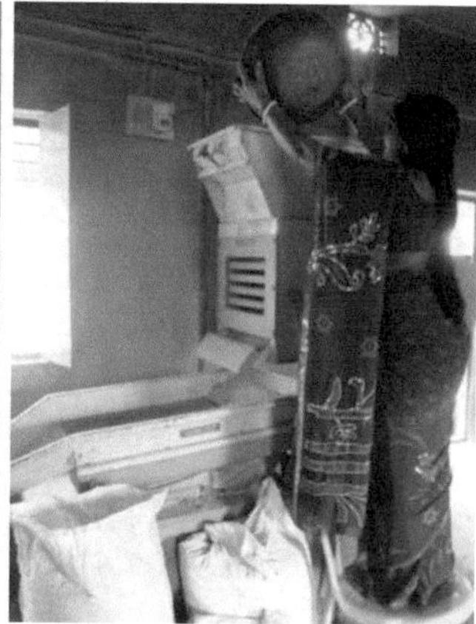

Fig. 10.6 'Nutrimix' making by members of MDMMS

There is irregular electricity supply in this area, and the quantity and quality of the product suffer due to it. The two-stage biomass gasifier will provide regular and clean supply of electricity at a low cost, which will increase the production and their income by over 500%. The increase in production will support more women and children at the Anganwadi centres, and the increase in the income generated will improve standard of living of the workers and give them an opportunity to explore more activities and build training centres.

The two-stage gasifier has already been installed and is running at the site as shown in Fig. 10.7. The project aims to supply the surplus electricity to nearby villages. The villages in the vicinity of project sites are electrified through utility grid but the availability of electricity in these villages is erratic and irregular, particularly during evening time. In order to supply the surplus electricity, two villages, namely Sindhitota and Karli, were identified and found technical feasibility to connect with biomass gasifier system using utility grid. A tariff petition has been submitted to OERC for permission of grid integration and tariff gap funding for household electrification through the gasifier. This involves obtaining regulatory approval from OERC for project-specific tariff for electricity generation from biomass gasifier, approve the electricity generated by the biomass gasifier to be supplied to the village using distribution assets of the DISCOM to be considered as deemed supply of the DISCOM and arrangement to allow interconnection of gasifier supply system to LT side of distribution transformer in a safe and secure manner. The project team is currently in dialogue with different stakeholders and process of completing the necessary formalities.

The gasifier is able to reduce the fuelwood consumption in grain drying by 65 kg/h with monthly saving of Rs. 3250 per month. The batch time is also reduced by 25% because of better heat supplied through biomass gasifier. Indoor air quality has noticeably improved but not quantified. The MDMMS is now planning to expand the production capacity by adding *Haldi* (turmeric) processing unit in their facility and increased quota for processing nutrition mix production.

Fig. 10.7 Two-stage biomass gasifier in Rayagada, Odisha

10.10 Recommendation for Future Research and Application in Practice

Universal electricity access needs a rural energy services model that should be for lighting/consumption plus economic/productive loads and compatible or integrated with grid. After the first successful implementation of the two-stage biomass gasifier in Rayagada, Odisha, three more are being implemented in the selected sites in both grid and off-grid areas along with viability gap funding by the state government. The project replicates installation of two-stage biomass gasifier in four remote areas in the states of Odisha and Madhya Pradesh in India. This is a pilot programme that will be the pathway to scale up this methodology to implement it in other parts of India and other countries. Monitoring and impact evaluation will be done for the technology as well as of the business model of the livelihood activity. Learning from these four pilot implementation projects, this model should be scaled up for more villages in India to support income-generating activities and improved electricity supply, specifically for women and provide energy access to households.

Second, the existing off-grid biomass projects establish to interface with the utility grid, it will considerably increase the viability and sustainability and further improve the electricity access for the households and livelihood activities in the rural areas. The project has already taken up policy work related to creating synergy between utility grid supply and biomass gasifier based DDG systems at existing village level.

The concept is well received by OREDA and is now also being planned for SPV-based electrification projects implemented by OREDA in the state of Odisha.

There is a need for policy and regulatory interventions of the state government to ensure integration of decentralized renewable energy generation/distribution with mainstream electricity distribution and skills development programme to include biomass energy for developing a cadre of local entrepreneurs for operation and management.

The fiscal incentives available under various schemes under different ministries of the Government of India can help achieve large-scale implementation of such models. The Ministry of New and Renewable Energy recently launched a programme to promote implementation of projects on renewable energy for rural livelihoods with provision of up to 65% investment subsidy to cover the renewable energy system cost. The Ministry of Micro, Small and Medium Enterprises is operating a scheme namely 'Credit Linked Capital Subsidy Scheme (CLCSS)' for technology upgradation of micro- and small enterprises. The scheme aims at facilitating technology upgradation of micro- and small enterprises (MSEs) by providing 15% capital subsidy (limited to maximum Rs. 15 lakhs) for purchase of plant and machinery.

The installation of the two-stage biomass gasifier has a huge potential and need for scaling up and replication in other sectors and also abroad. This technology with our existing business model of targeting rural livelihood clusters and household electrification can be beneficial for many developing countries, such as Myanmar, Nepal, Bangladesh, Kenya, Burkina Faso and other African countries. It can also be used purely for rural electrification or captive livelihood clusters.

Annexure 1

Livelihood clusters in which scoping study was done

S. No.	Cluster activity	NGO/agency involved	Location
1.	Nutrimix making	AKSUSS	Rayagada, Odisha
2.	Cashew processing	SPREAD	Koraput, Odisha
3.	Cashew processing	Vasundhara	Nayagarh, Odisha
4.	Red chilli processing	Vasundhara	Sambalpur, Odisha
5.	Nutrimix making	ORMAS	Puri, Odisha
6.	Bamboo products making	ORMAS	Puri, Odisha
7.	Rural electrification	Forest Department and CPRD	Mayurbhanj, Odisha
8.	Rural electrification	Forest Department and DEEP	Keonjhar, Odisha

S. No.	Cluster activity	NGO/agency involved	Location
9.	Lemongrass oil distillation	KGVK	Ranchi, Jharkhand
10.	Stone crushing	KGVK	Ranchi, Jharkhand
11.	Tussar silk reeling	Jharcraft	Saraikela-Kharsawan, Jharkhand
12.	Chiwda making	Women Line	Dumka, Jharkhand
13.	Nutrimix making	Women Line	Dumka, Jharkhand
14.	Bamboo basket making	Women Line	Dumka, Jharkhand
15.	Lift irrigation	Samvad	Deoghar, Jharkhand
16.	School and hostel	Sarva Shiksha Abhiyan	Ranchi, Jharkhand
17.	Milk chilling	MPDPIP	Shivpuri, Madhya Pradesh
18.	Poultry feed making	MPDPIP	Sidhi, Madhya Pradesh
19.	Agro processing	CIAE & MPVS	Chhindwara, Madhya Pradesh
20.	Bamboo furniture making	MPSBM	Balaghat, Madhya Pradesh
21.	Bamboo furniture	MPSBM	Harda, Madhya Pradesh

References

Dhingra, S. & Tanvir, B. (2014, October 2). Electrification through two-stage biomass gasifier: Opening livelihood opportunities for rural communities in India. *Akshay Urja* (Magazine by Ministry of New and Renewable Energy, Government of India) 8, 26–27.

Ministry of New and Renewable Energy, Government of India. (2015). *Tentative State-Wise Break-Up of Renewable Power Target to be Achieved by the Year 2022*. Available from: http://mnre.gov.in/file-manager/UserFiles/Tentative-State-wise-break-up-of-Renewable-Power-by-2022.pdf. Accessed November 6, 2015.

World Bank. (2011). India: Biomass for sustainable development; lessons for decentralized energy delivery; village energy security programme.

Part V
Sustainable Habitat

Chapter 11
MiraMap: A Collective Awareness Platform to Support Open Policy-Making and the Integration of the Citizens' Perspective in Urban Planning and Governance

Francesca De Filippi, Cristina Coscia and Roberta Guido

11.1 Introduction

When exploring innovative approaches for a more inclusive and sustainable urban planning and governance using the ICTs, the set up of the methodological framework is particularly relevant in order to address the complexity and dynamics of urban development, and to deal with the interaction of multidisciplinary concepts and contributions, as it will be demonstrated through the case study in Torino (Italy).

MiraMap is an ongoing project led by the Politecnico di Torino (Italy), deeply rooted in a pilot experience named Crowdmapping Mirafiori Sud (CMMS): the aim is to set up a governing tool which integrates citizens' perspective—through their effective engagement—in the design and production of public services and the use of a collaborative platform, which benefits from a social networking and a web-based mapping system.

Thus, the project takes into account both the application of participative methods and techniques, which support the community to identify problems and resources, and the integration of data and development of ICT-governing tools for public stakeholders. Participative planning is then intended as a way to think over the public action, either in the relationship with citizens or in the public space management. Moreover, the integration of eGovernment and social network paradigms is experimented here to enlarge the target of users and, in doing so, fostering citizen engagement and empowerment.

F. De Filippi (✉) · C. Coscia
Department of Architecture and Design, Politecnico di Torino, Turin, Italy
e-mail: francesca.defilippi@polito.it

C. Coscia
e-mail: cristina.coscia@polito.it

R. Guido
Department of Architecture Design and Urban Planning, University of Sassari, Sassari, Italy
e-mail: rguido@uniss.it

S. Hostettler et al. (eds.), *Technologies for Development*,
https://doi.org/10.1007/978-3-319-91068-0_11

11.2 International Policies and Action Plans on Civic Engagement and Social Innovation in Urban Development

The rising people's demand to have access and to be involved in decisions dealing with their own urban context led European Union (EU) institutions to consider the role of participation for its policies. Although urban planning is not a specific European policy, competence, actually economic, social and territorial cohesion all have a strong urban dimension: as the vast majority of Europeans live in or depend on cities, their developments cannot be isolated from a wider European policy framework (European Commission, DG Regional Policy 2011b).

The URBACT III Operational Programme 2014–2020 includes among its specific objectives to ensure a participatory approach through the involvement of the relevant stakeholders in the action-planning process (URBACT Study 2015; URBACT II 2015) .

Also at international level, the United Nations Agenda (Transforming our world: the 2030 Agenda for Sustainable Development) includes participation among post-millennial Goals. Indeed, one of them (16) aims to 'Promote peaceful and inclusive societies for sustainable development, provide access to justice for all and build effective, accountable and inclusive institutions at all levels'—together with its commitment (16.7) to 'responsive, inclusive, participatory and representative decision-making at all levels'.

Furthermore, it cannot be avoided to consider the arising of platforms, software and applications, often seen as solutions to societal needs: they enable exchange, communication and the creation of a community of citizens and other stakeholders on shared interests and concerns. They are seen as tools to empower citizens, including marginalized groups, improve public services and at the same time ensure equal access to information and promote democracy (European Commission 2015).

For this reason, EU bodies began recognizing the role of ICT to foster new forms of civic engagement in urban planning, as a social innovation and to support social innovations.

Within the *Europe* 2020 *Strategy*, the *Digital Agenda* recognizes the key role of public administrations in creating the condition to foster social innovation, becoming more and more 'open, flexible and collaborative in their relations with citizens' and promoting the 'eGovernment to increase their efficiency and effectiveness and to constantly improve public services in a way that caters for user's different needs'. At a time of highly constrained public resources, ICT is seen as a tool to help the public sector develop innovative ways of delivering its services to citizens while unleashing efficiencies and driving down costs.

Within the *2020 Digital Agenda*, it has been launched the European eGovernment Action Plan 2011–2015—harnessing ICT to promote smart, sustainable and innovative Government [COM(2010) 743 final]. Based on the Malmö Ministerial Declaration of 2009, it sets out the objectives for public administrations to invite third parties to collaborate on the development of eGovernment services, strengthen the transparency of administrative processes and involve stakeholders in public policy processes.

In particular, its priority *User empowerment* stresses the importance of increasing the capacity of citizens and organizations, promoting the development of services designed around users' needs, and inclusive services, the collaborative production of services, the re-use public sector information, improving transparency and fostering the involvement of citizens and businesses in policy-making processes.

The Action Plan underlines that 'social networking and collaborative tools enable users to play an active role in the design and production of public services' (eGovernment Action Plan 2011–2015, 2.1.2). However, it invites to explore which are the most suitable tools and how best to apply these to effectively engage civil society and individual citizens.

Also, the *European Innovation Partnership on Smart Cities and Communities* (establishing strategic partnerships at the local level and across borders in Europe) recommends new tools of engagement. One of the main actions proposed is to *'implement collaborative, integrated smart city planning and operation, that maximise city-wide data to deliver more agile processes; employing modern multi-criteria simulation and visualisation tools' (EIP Smart Cities and Communities,* Strategic Implementation Plan 2013).

The focus of innovation's needs for Europe was defined in the European research programme Horizon 2020, that also addresses funding to projects that promote platforms to set up more participatory democratic processes and to support grassroots processes and practices to share knowledge. Collective Awareness Platforms are expected to have very concrete impacts to foster open democracy, open policy-making (better decision-making based on open data) or in new collaborative approaches to inclusion (*Horizon 2020 Work Programme 2016–2017, 5.i. Information and Communication Technologies*).

In the report *Cities of Tomorrow—Challenges, visions, ways forward* (European Commission, DG Regional Policy 2011b), the European Commission made the following recommendations for actions: empowering cities to define their own policies related to their context; ensuring transversality of policies and impact of one area on the other; supporting cities but leaving them room for *manoeuvre* in connecting with citizens; letting cities decide on their own priorities.

The Bureau of European Policy Advisors gave a definition of social innovation: 'Innovations that are social in both their ends and in their means. Specifically, we define social innovation as new ideas (products, services and models) that simultaneously meets social needs and creates new social relationships or collaborations. In other words, they are innovations that are not only good for society but also enhance society's capacity to act' (BEPA 2011a, p. 33).

Social innovation is, therefore, considered at the heart of reshaping society: it can be used and developed, both as a means and as an end to city governance. However, the bridge between these diffused initiatives and the ability to catalyze them into inclusive governance is often missing, so, it is solicited a more integrated connection (De Filippi et al. 2017).

11.2.1 Citizens Participation Through the ICTs: The Global Scenario

An increasing demand from citizens to participate and collaborate to the future urban scenarios, especially at local and regional scale, have also challenged democracies all over the world (Held 2006). Public administration of representative democracies have, thus, progressively adopted policy frameworks to become more responsive by taking more participatory elements over and by opening up to the public in many fields. As a result, it has involved a virtuous mechanism by which political framework of public administration drives social innovation by promoting bottom-up approaches to policy-making for better governance and sustainable development (Horita et al. 2015; Davies et al. 2012).

The development of ICTs (especially, user-driven applications) has widely been recognized as a way of encouraging communications between people by transforming the way they interact and they use the Internet (Ratti 2013).

Web-based services are excellent opportunities to improve three broad qualities of good governance like enhance transparency, people participation and public services in a way more cost-effective and accountable for citizens (Innes and Booker 2004).

ICT tools for eGovernment can enhance public engagement and permit a wider percentage of the population to contribute to the public management. ICTs thereby are seen as tools to better enable participation, democracy and more inclusive societies, evolving from traditional top-down hierarchical models towards networked models, to facilitate interactions between urban stakeholders and actors (Silva 2010).

However, a number of critical issues and challenges still need to be tackled. Many of them can be related back to the lack of skills and to the shortcomings of both the ICT-enabled tools; moreover, the digital illiteracy and the digital divide. These effects endure both in the Global North and in the Global South between elders and generations 'born digital' as between urban centres and peripheries or rural areas; the question requires to be put forward concerning the Global South, in which too often applications of urban planning, eGovernment, ICTs only partially address the real challenges facing sustainability (Priti 2006). It is because models are built in and for the North and then transferred to the South, without having been replanned to the specific objectives, but simply adapted (Bolay 2015).

Nevertheless, ICT performance will remain crucial not only in the Global North countries for sustaining long-term development and enhancing governance models, but also in the emerging and developing countries in fostering structural transformations, increasing efficiency as well as reducing the digital, economic, and social divides within their territories (World Commission on Environment and Development 1987; World Conservation Union 2006; WSIS, World Summit on the Information Society 2003).

11.3 Related Work on the Field

The MiraMap project in Torino has a common goal with other systems, offering an online platform which allows citizens to interact with the public administration and to send information (De Filippi et al. 2017b). They all have a transparent interface and are easy to use, and allow to see the warning list and to check the status.

Differently than other projects based on maps such as the successful open-source solution FixMyStreet in UK, a platform where people can send information and discuss local problems about infrastructural issues, or the commercial products ePart and ImproveMyCity, MiraMap focuses also on the proactive part where citizens can report proposals and positive aspects of their neighbourhood. Differently than other proposals such as IRIS Beta in Italy, which have a ‘social network’ character, it is based on an interactive map, which multiplies the visibility of citizens actions. Other community-aimed solutions such as Streetlife and EveryBlock do not connect citizens with the administration.

Regarding web platforms and applications developed for residents of a neighbourhood or specific locality, we can differentiate three generations of technologies on the basis of their interactivity (De Filippi et al. 2016).

First, we can consider the numerous community portals which list local businesses and services, and are produced often by local residents, such as through user-generated content, ranging from news to event listings (e.g. www.lovecamden.org, http://www.sansalvario.org/, etc.). Even if the content is shown in web pages without the use of maps, the geographical nature of the information shared change, becoming based on specific areas of the cities, such as neighbourhoods. Thus, the general objective is to provide online information to those who are interested in getting to know what happens in a given part of the city.

Second, a recent approach has gone a step beyond information provision by enabling people to have a direct link to others who live around them. Sometimes people are also supported to engage with local businesses, associations and/or governing bodies. Examples of such approaches include the EU-funded MyNeighbourhood platform (www.my-n.eu/da) and the Polly & Bob platform in Germany (blog.pollyandbob.com/). Discussions are enabled by blogs, discussion forums, event calendars, etc. In this case, simple GeoWeb applications enable citizens to map POI and events. The general thrust is to encourage people to get involved within their own neighbourhoods and engage their family and friends to do the same. Data and functionality of existing City Information Apps (e.g., MyCityWay, Foursquare) are combined with tools that connect people locally. My Neigbourhood also experiments with basic gamification techniques to stimulate community building.

Whether in the first case the approach was mainly based on information, here the focus is on facilitating communication between people.

Third, the applications of open-source software in post-emergency situations, such as Ushahidi, that has been adopted—as will be demonstrated—in the pilot Crowdmapping Mirafiori Sud/CMMS in Torino. Ushahidi, developed in Kenya to map in 2008 the violence in the post-electoral period, is an open-source platform, which allows an easy crowdsourcing of data and the total transparency of their diffusion (Hagen 2011). Ushahidi is nowadays used as a prototype and an example of something that could be done by matching information generated from citizens' reports, media and NGOs into a geographical map.

Finally, map-based services have been used to push the attention at problems or things that have to be changed in the cities. This generation of services has only indirectly involved Local Authorities, since interaction with the platform on the Institution's side is not allowed. It is worth mentioning Infalia—Improve My City and FixMyStreet where problems are reported on a map in order to be addressed by the local Council, but not directly managed. Another example is Changify platform (www.changify.org), which particularly focuses on people who wish to share things they love or would like to see changed in their neighbourhood.

Current online neighbourhood portals are, therefore, primarily directed towards strengthening community life with help of online technologies, thereby engaging citizens to communicate and discuss any issue of interest.

Considering MiraMap functionalities, it can be included within this third generation of technologies as well as FixMyStreet and Improve My City, but differently from them, it focuses also on the propositive part of citizens, who can report proposals and positive aspects of their neighbourhood. It aims at further increasing engagement and at promoting co-production of services by means of the social networking functionalities (Kingston 2007).

11.4 From a Pilot to a Governing Tool: A Case Study in Torino (Italy)

As mentioned in the Introduction, the case study in Torino has been developed in two phases: first, the application of participation techniques on a simplified prototype (Crowdmapping Mirafiori Sud/CMMS) and, second, the development of a more sophisticated IT solution (MiraMap).

The first phase carried out in 2013 in a bound and determined area of the Mirafiori Sud district, has been addressed to investigate if the use of the ICTs could be a means to foster social inclusion.

The second phase, in 2015, implemented in the same reference context but in a wider geographical area, develops a more structured approach both in terms of IT system and governance model. The participatory approach developed during the first phase has been essential to validate and to foster technological benefits.

Fig. 11.1 Public meetings with the engaged stakeholders

11.4.1 The Methodology

Phase 1
The overall goal of the CMMS pilot experience has been to establish strong connections within the territory and institutional relations with the local actors. The crowdmapping process implemented in this phase has been useful not only to sensitize the population and to define the state of the art, but mostly to analyse and share the results together with all the stakeholders engaged (the community, the public administration and the researchers) in order to hypothesize active and participative solutions (Fig. 11.1).

The project involved the academic (including students) and the local community in a participative and inclusive process to identify and report on a web-based map the obstacles/barriers—either them being physical, spatial or cultural—which prevent vulnerable categories to access and use the public space in the neighbourhood. In order to allow an easy crowdsourcing of data and the total transparency of their diffusion, the open-source platform 'Ushahidi' has been adopted and customized. The technological tool provided a free database to gather collective information and show them on a web-based crowdmap (Fig. 11.2).

One of the key elements offered by Crowdmap-Ushahidi is the use of mobile phones as a way to report and receive updates, not needing an Internet connection,

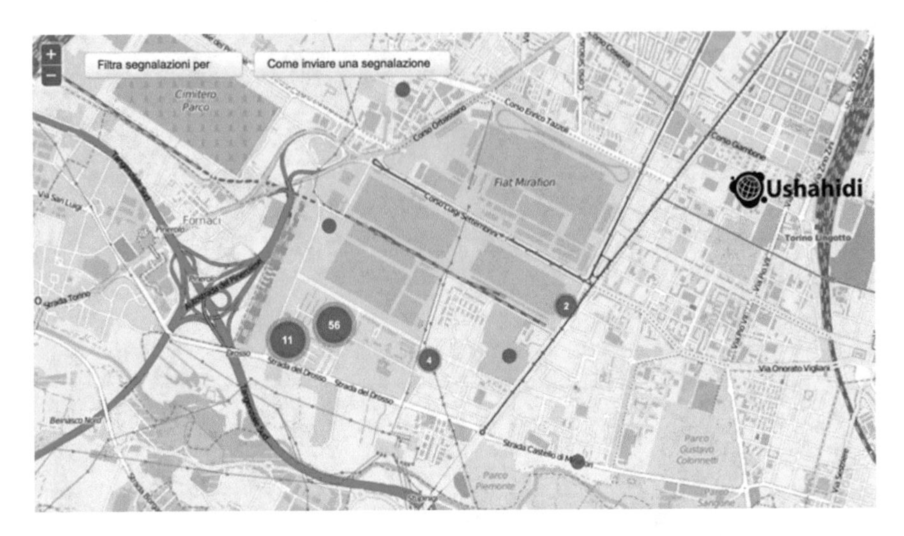

Fig. 11.2 Crowdmapping reports on the web-based map provided by Ushahidi

which might not always be available. That was an essential element for the implementation of the project in order to achieve social inclusiveness by providing the possibility to report from any kind of mobile device. Outcomes of data collection were published to make them accessible to the local authorities. In the meanwhile, an analysis of data was needed, in order to understand the weak points and to further discuss with people, and so a plan of activities, such as traditional meetings to transect walks, was set up (De Filippi et al. 2017a). This process was important to enhance public participation, involving people from the first to the last step (Fig. 11.3).

Phase 2

The results achieved have been essential to move to the next phase. The MiraMap platform has been set up in order to answer requests from citizens and stakeholders. Their positive feedback has been considered as an essential prerequisite to design a proper governing tool facilitating their effective engagement.

The process expects to involve both citizens and the local administration in a report process of critical issues as well as positive trends and resources within the Mirafiori Sud district area, throughout the use of a digital platform made up of a geo-referenced map combined with a BPM—Business Process Management system (De Filippi et al. 2016).

The interactive map is used by citizens to report problems and proposals in the neighbourhood, making them visible to everyone. The BPM is used by the administrative staff to manage the reports and give feedbacks. The map automatically shows the progress of the administrative process as the workflow proceeds in the BPM, and it provides citizens and policymakers with a comprehensive view of problems and opportunities of the neighbourhood (Fig. 11.4).

The regular stakeholder engagement at different stages of the ICTs toolset and development process helped in specifying and validating necessary and common

Fig. 11.3 Data collection with the community (transect walks)

requirements specification. In particular, in order to ensure the tool compliance and integration into the current administrative process, the managing executives and the public officers have been involved in each step in co-creating and testing the technological platform.

At the same time, a monitoring and evaluation process, based on the Community Impact Analysis (CIA) method (Lichfield 1996; Coscia and De Filippi 2016) has been defined and set up.

Fig. 11.4 The MiraMap website

11.5 Achievements and Further Research Steps

In the perspective of enhancing strategies, approaches and tools for driving innovation in urban planning, management and governance at different scales and generating social impact through the use of the ICTs, the specific case study aims to give a contribution in terms of lessons learned. In particular, some key issues should be tackled and achieved in the planning process:

1. the design of a collaborative platform and a methodology able to foster social inclusion and innovation, starting from an accurate analysis of residents' needs;
2. the integration of the technology (online) with the participatory process (offline), to enhance social impacts strategies and promote community empowerment;
3. the compliance and integration of the tool into the administrative process (workflow), promoting the openness, the transparency and the accountability of the local government;
4. the capacity building process, involving both public officers, administrators and the community;
5. the setup of a monitoring and evaluation process;
6. the development of a strategic and action plan to support the replication and scale-up of the piloted action.

The tool makes possible for the community to be involved in co-designing and co-producing services (De Filippi et al. 2017b). From the Local Government side, having enabled it to access and produce data, it builds and strengthens its accountability. The workflow management needs to suit as best as possible to features of the administrative process in use and, above all, to become an opportunity to make it more efficient thanks to the methodology adopted to get feedback from administrative staff, that is made by an iteration of testing phases after fast IT developments.

In 2016, Torino was awarded consistent funding by the Italian government in order to implement innovative regeneration projects in 'peripheral' areas. The inclusion of the MiraMap project among the actions of the city opens up further opportunities for work on the platform.

The next step (third phase) regards the replicability of the method and the model in other administrative areas and scalability to the metropolitan scale, with a substantial commitment to expanding interoperability with other administrative tools and in communicating data to citizens. Further analysis will be dedicated to assessing the social impact and the effectiveness of the process.

Collaborative platforms, starting with their experimentation in marginalized areas, can provide a good field for reflection on how the involvement of citizens can be stimulated through innovative public policies which aim to assess the value of common assets and to structure their shared management.

Acknowledgements The technological platform for MiraMap (second phase) has been developed in collaboration with the Social Computing team of the University of Torino coordinated by Prof. Guido Boella. First Life (www.firstlife.org) is a project awarded to the above-mentioned team and funded by the call Smart Cities and Communities and Social innovation of Ministry of University and Research of Italy.
The authors thank A. Cantini who helped with the preparation of the paper, while she was participating in the project, namely in the Academic Year 2015/2016.

References

Books

European Commission. (2011a). *Empowering people, driving change: Social innovation in the European Union.* Bureau of European Policy Advisors (BEPA).
European Commission. (2011b). *Cities of tomorrow—Challenges, visions, ways forward.* DG Regional Policy.
European Commission. (2015). *Horizon 2020 work programme 2016–2017, 5.i. Information and Communication Technologies.*
Held, D. (2006). *Models of democracy.* Stanford, Redwood City, CA: Stanford University Press.
Horita, M., et al. (Eds.). (2015). *ICT tools in urban regeneration.* New York: Springer.
Lichfield, N. (1996). *Community impact evaluation.* London: UCL Press.
Ratti, C. (2013). *Smart city, smart citizen. Meet the media guru.* Milano: EGEA.
Silva, C. (2010). *Handbook of research on E-planning: ICTs for urban development and monitoring.* Pennsylvania: IGI Global Snippet.

URBACT Study. (2015). *New concepts and tools for sustainable urban development 2014–2020, synthesis report*.
United Nations. (2015). *Transforming our world: The 2030 agenda for sustainable development.*

Journal Articles

Bolay, J. C. (2015). Urban planning in Africa: Which alternative for poor cities? The case of Koudougou in Burkina Faso. *Current Urban Studies, 3,* 413–431.
Coscia, C., & De Filippi, F. (2016). The use of collaborative digital platforms in the perspective of shared administration. The MiraMap project in Turin (EN version). *Territorio Italia, 1,* 61–104.
Davies, R. S., Selin, C., Gano, G., & Pereira, G. Â. (2012). Citizen engagement and urban change: Three case studies of material deliberation. *Cities Elsevier Journal, 29*(6), 351–357.
De Filippi, F., Coscia, C., Boella, G., Antonini, A., Calafiore, A., Guido, R., et al. (2016). MiraMap: A we-government tool for smart peripheries in smart cities. *IEEE Access, 4,* 3824–3843.
De Filippi, F., Coscia, C., & Cocina, G. (2017a). Piattaforme collaborative per progetti di innovazione sociale. Il caso Miramap a Torino, *Techne* n. 14 (2017). Firenze University Press.
De Filippi, F., Coscia, C., & Guido, R. (2017b). How technologies can enhance open policy making and citizen-responsive urban planning: MiraMap—A governing tool for the Mirafiori Sud District in Turin. *International Journal of E-Planning and Research (IJEPR), 6*, https://doi.org/10.4018/ijepr.
Innes, J. E., & Booher, D. E. (2004). Reframing public participation: Strategies for the 21st century. *Planning Theory and Practice, 5*(4), 419–436.
Kingston, R. (2007). Public participation in local policy decision-making: The role of web-based mapping. *The Cartographic Journal, ICA Special Issue, 44*(2), 138–144.
Priti, J. (2006). Empowering Africa's development using ICT in a knowledge management approach. *The Electronic Library, Emerald Group Publishing Limited, 24*(1), 51–67.

Online Publications

Communication from the Commission to the European Parliament, the Council, the Economic and Social Committee and the Committee of the Regions of 15 December 2010—The European eGovernment Action Plan 2011–2015—Harnessing ICT to promote smart, sustainable and innovative Government [COM(2010) 743 final]. http://eur-lex.europa.eu/legal-content/EN/TXT/?uri=celex:52010DC0743. Accessed November 14, 2017.
EUROPE 2020 A strategy for smart, sustainable and inclusive growth [COM(2010)2020 final]. http://eur-lex.europa.eu/legal-content/EN/TXT/?uri=celex%3A52010DC2020. Accessed November 14, 2017.
European Innovation Partnership on Smart Cities and Communities—Strategic Implementation Plan. (2013). http://ec.europa.eu/eip/smartcities/files/sip_final_en.pdf. Accessed November 14, 2017.
Hagen, E. (2011). Mapping change: Community information empowerment in Kibera. *Innovations*, *6*(1), 69–94. http://mapkibera.org/wiki/images/4/42/INNOVATIONS-6-1_Hagen.pdf. Accessed November 14, 2017.
URBACT II. (2015). *Capitalisation. State of art, social innovation in cities*. http://urbact.eu/capitalisation2015/catalogue/social/appli.html?summary=complex. Accessed November 14, 2017.

World Commission on Environment and Development. (1987). *Our common future, from one earth to one world.* http://www.un-documents.net/our-common-future.pdf. Accessed November 14, 2017.

World Conservation Union. (2006). *The future of sustainability*, https://www.iucn.org/. Accessed November 14, 2017.

WSIS, World Summit on the Information Society. (2003). http://www.itu.int/net/wsis/. Accessed November 14, 2017.

Chapter 12
Reaching the Last Mile—Technology Solutions and Models for Service Delivery

Sanghamitra Mishra and Koneru Vijaya Lakshmi

12.1 Background

About 884 million people worldwide do not have access to safe drinking water and are, therefore, forced to consume water that has high concentrations of bacteria, viruses, protozoa or even chemical contaminants. As a result, more than 4000 children die every day from diarrhoea (WHO 2011). 80% of the water-borne diseases in developing countries originate from microbiological contamination.

Over the past few years, private market actors such as Hindustan Unilever Ltd., Tata Chemicals, PATH, Aquatabs, Antenna Technologies, Procter & Gamble, TARA water services, Water Health International, Naandi have come up with promising technological solutions such as silver-impregnated nano-technology-based water filters, membrane filters, chlorine solution and tablets, water kiosks based on UV and membrane filters, slow sand filters, etc., for low-income customers. It is widely recognised that new and emerging technologies such as nano-based water filtration, have the potential to service this population due to their high efficacy, low-material intensity and cost (Development Alternatives, May 2011).

Despite the availability of various technologies in the market to deal with the water quality, the benefits have not percolated down to the 835 million strong BoP population in India (Vijaya Lakshmi et al. 2011; 2013), who live on an annual income of less than INR 200,000 (<US$3000). It becomes imperative to offer them affordable yet effective solutions for water purification. These solutions often do not reach the BoP segment due to high cost of the filters, lack of awareness and demand for safe water and irregular supply of products and after-sales services. Thus, any innovation

S. Mishra
Society for Development Alternatives, Currently in Oxfam India, New Delhi, India
e-mail: misra2011sanghamitra@gmail.com

K. Vijaya Lakshmi (✉)
Society for Development Alternatives, New Delhi, India
e-mail: kvijayalakshmi@devalt.org

S. Hostettler et al. (eds.), *Technologies for Development*,
https://doi.org/10.1007/978-3-319-91068-0_12

needs to address these issues in order to ensure adoption and sustained demand. To ensure the long-term viability and sustainability of these models, it is essential to develop them in consideration with the socio-economic status of the community and market conditions in the area including the study of the flow of finance in those markets.

There is sporadic evidence that technology research benefits can reach the BoP, through innovative and appropriate delivery models (TARA 2012). In view of the same, Development Alternatives (DA) Group has piloted an action research for ensuring access to safe water to the BoP through testing innovative delivery models using nano-technology-based filters. The action research looked into **testing its scalability potential** and associated barriers of these models for a wider dissemination of these products in BoP market.

12.2 Piloting the Delivery Models

12.2.1 Technology Selection Process

Based on the readiness to be taken to the market in terms of economics and availability of product, an initial scan resulted in the short listing of technologies for the pilot. Five technologies made the cut based on performance, ease of use and wow factors. It was decided to include the end-user in the final selection since it was a market-based model. During the consultations with target end users, the filters and their associated features were demonstrated to the residents. The feedback from the group helped analyse and rank the selection criteria for the users. These are categorised as critical and important factors in decision-making for the residents are listed in Table 12.1.

The decision in Table 12.2 shows the mapping of various technology options against the above factors. The possible lists of three filters tested from the matrix are

Table 12.1 Critical, important and wow factors influencing decision-making for purchasing water filters

Critical factors	Important factors	Wow factors
Attached storage unit	Rate of filtration	Aesthetics
No electricity	Ease of installation and up-keep	Tamper-proof mechanism
Purifies and changes taste and smell	Sturdiness of filter tap	
Sturdy and durable material	Familiarity of shape/design	
Cost of ownership		
Service support		

Table 12.2 Factors considered for filter selection

Factors/filters		Tata Swach	Life straw 1	Life straw 2	Tulip Siphon	ARCI candle
Critical	Attached storage unit	✓	×	×	×	✓
	No electricity required	✓	✓	✓	✓	✓
	Purifies and changes taste and smell	✓	×	✓	✓	✓
	Sturdy and durable material	✓	×	×	×	✓
	Cost of ownership	✓	✓	×	✓	✓
	Service support	✓	✓	✓	×	×
Important	Rate of filtration	✓	✓	✓	✓	×
	Ease of Installation	✓	×	×	×	✓
	Sturdiness of filter tap	✓	✓	×	×	✓
	Familiarity of shape/design	✓	×	×	✓	✓
Wow	Aesthetics	✓	×	×	✓	✓
	Tamper-proof mechanism	✓	✓	✓	×	×

(a) **Swach Smart filter**
(b) **Tulip Siphon filter**
(c) **ARCI candles with container model filter**.

On the overall product deliverables across all factors, service capabilities and probability of product continuity, Tata Swach Smart filter was selected for pilot testing in Delhi and Bundelkhand.

12.2.2 Demand Creation for Safe Water in the Study Area

As anticipated, there was a lack of awareness of the need for safe water in many households. Even where there was recognition for the need of safe water, it rarely converted into a priority due to a slew of reasons ranging from competing priorities, to the lack of awareness about options or the lack of acceptance that their water was not safe. The perception of safe water is closely tied to appearance, taste and smell. The families who opted for safe water, they bought water from the local RO vendor. In order to encourage behaviour change, several promotional activities were conducted using street plays, door-to-door visits, leaflets, posters and banners containing information on the possible impacts of drinking contaminated water and benefits of using Swach filters.

12.2.3 Innovations in Delivery Model

12.2.3.1 Lead Experience User (LEU) Model in Delhi

Bacterial contamination is one of the largest problems in urban slums in India. The quality of water supplied or sourced in these areas is secondary contamination due to unsanitary and unhygienic living conditions. Delhi is no stranger to this problem (Kumar 2013).

DA has developed an innovative approach to reach the BoP in two urban slums of Delhi, i.e. Mandanpur Khadar and Ambedkar Camp. The technology selected for the pilot was nano-silver-based Swach filters. A peer influence-based model was adopted. The pilot was implemented in two slums, Madanpur Khadar in South East Delhi (15,000 HHs) and Ambedkar Camp in South West Delhi (3000 HHs). The average monthly income of a household in both the slums vary from Rs. 5000 to Rs. 10,000. In Madanpur Khadar, the primary source of water is hand pumps, used mostly for cooking and washing. Some families also use it for drinking. Majority of the residents rely on the so-called filtered water or packaged mineral water from the local RO plants in the area. These residents are provided with door-to-door services at prices as low as Rs. 2 per litre of RO water. Ambedkar Camp residents mainly rely on water tankers supplied by the Delhi Jal Board (the city water supply system). The main concern for water quality in these areas was bacterial contamination.

Service Delivery Model

A LEU is a local housewife willing to be a Swach Champion. Swach is presently being sold based on two distinct distribution models, i.e. direct selling to local Swach champions and the distributor model shown in Fig. 12.1. Based on the local market conditions and taking into account the existing distribution channels for Swach, a LEU model was adopted. The model builds on the existing distribution channels but enables last mile delivery to a previously sub-serviced population. The LEU cuts through the long-winded distribution chain and obtains filters at a discount to ensure access of the slum population to a filter. This modified delivery model was chosen for the pilot due to a number of different reasons. First, there was no requirement of working capital to implement the model. Second, there was already a linkage with the existing distributor of Tata and local delivery by these distributors. In addition, this model allowed for direct activation and training by Tata Chemicals as well as access to free servicing by Swach service executives.

The two elements of the model which make it distinct from the existing distributor model, is access to a local LEU and the use of local retailers to stock and sell replacement units and bulbs were also important reasons for this model being chosen. The model considers local retailers to distribute the replacement filter units/bulbs as these agents have their own pre-existing relationships with clients, which they are able to leverage in order to make sales. Second, it requires little to no working capital and the margins will be suitably high as compared to the other products that they stock, making them a good investment.

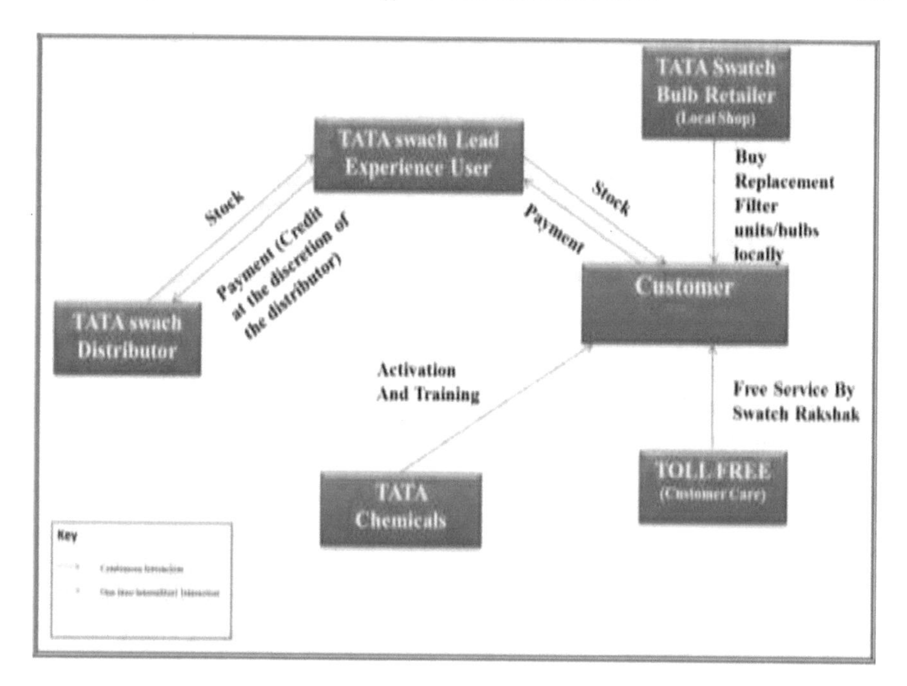

Fig. 12.1 LEU model schematic for Tata Swach

Financing was a critical element to consider, given the low-income levels of the BoP population. Following the research of the pilot area however, the cost of the filter at Rs. 900, was not considered to be prohibitively expensive and would not deter residents from purchasing the product. Although it was understood that providing financing options would increase the rate at which the filters were adopted, as there were no financing organisations working in that area, the decision to provide financing as part of the model was not built into the model.

On the Ground Activities

In order to reach the maximum number of households in the study areas and to test the workability and potential scalability of the model, following processes were taken up.

(a) **Selection and Activation of Lead Experience Users (LEUs)**: LEUs are local women/housewives who are interested in becoming entrepreneurs. Given the socio-cultural environment in villages, LEUs are a vital component as they have good knowledge of the local people and existing networks in place, credibility as a user, and thus ability to demonstrate the long-term benefits of using the filter.

(b) **Training sessions** were conducted for LEUs to orient them to the product and its features. This is also essential for effective communication of the services being offered to the community. Initial on-field support was given to LEUs by experienced marketing professionals. The idea was to assist the SHG members so that they are confident to continue the activity in long term.

(c) **Financial incentives** were offered to the LEUs, so they could create a sustainable business from the activity. LEUs were able to purchase the filters from Tata Swach for the reduced cost of Rs. 799. The filter was then sold at the market price of Rs. 899 to the customers, leaving the LEUs with a profit of Rs. 100.

(d) **On-field Promotion and Demonstration**: A separate team of demonstrators was recruited for running live group demonstrations on the streets outside the homes.

(e) **Delivery and Post Delivery**: A daily tracking system was developed to track record of houses visited, hot leads identified and sales of the filter. The date of booking was recorded to help the LEUs determine the date of bulb replacement. Customers were given a toll free number of Swach filter in case of any issues.

Lessons Learnt

- **Need for Financial Incentives**: As per our initial research, there was no apparent requirement of financial incentives as the product fell in an affordable range for the consumers in urban slums. However, it was realised that affordability never ensured acceptance and adoption of the product. In some cases, they were used as small ticket payments (INR 5 for 10 L of water delivered at home) for their daily quota of water from a local RO supplier. Consumers preferred making daily payments to purchase water daily instead of paying a bulk amount for a filter which proved to be expensive in the long run with compromised water quality. Thus, in order to encourage adoption of water filters, the need for financial incentives was felt. But the absence of microfinance institutions in the study area, however, prevented this intervention.
- **Technologies need to be customised**: While the purification technology is not an aspect that has a major bearing on the consumers decision-making, the term RO (reverse osmosis) has become a household name. It has acquired a brand value people associate with purity, even though it may not be a reputed company or product. There are local vendors riding this wave and supplying sub-standard water to communities, as seen in Madanpur Khadar. In addition, the raw water TDS levels in Madanpur Khadar are high and have an impact on taste, even though they are well within permissible limits as per the BIS standards. Due to access to demineralized water (even though it may not be bacteria free) from the local RO plants people have acquired a different taste. The perception of safe water is so directly linked to taste that some families rejected the Swach filter on this account. This issue was not faced in Ambedkar Camp, where the raw water TDS levels were lower and did not distort taste.
- **Need for Behaviour Change Communication for Demanding Safe Water**: As expected, there was a lack of awareness on the need for safe water in many households. Even where there was recognition for the need of safe water, it rarely converted into practice due to a variety of reasons ranging from differential priorities, to the lack of awareness about options or the lack of acceptance that their

water was not safe. Also, the perception of safe water is closely tied to appearance, taste and smell. The families who opted for safe water, bought water from the local RO vendor. In order to encourage behaviour change, several promotional activities were conducted using street plays, door-to-door visits, leaflets, posters and banners containing information regarding the possible impacts on health by drinking contaminated water and highlighting key features of Swach filters.

- **Designing robust business models and entrepreneurship development for Supply Chains**: Discovering and nurturing entrepreneurship is not an easy task, more so in areas where women are not considered at par with men. Initial efforts of recruiting women as LEU, who have experience in promoting safe drinking water or similar schemes showed grim results due to their unavailability and time constraints to undertake additional work. Another major barrier was that most women were accustomed to the idea of a fixed monthly salary. Commission-based and target-driven work was outside their comfort zone. This can in part be attributed to the lack of confidence and risks involved in reaching out to communities and creating enough demand for water filters for the business to be profitable. They were unconvinced by arguments that this micro-entrepreneurship would provide long-term benefits. Thus, only two women eventually managed to take up the role of LEUs.
- Responding to the above challenge, the team widened the pool of people suited as LEUs including teachers, doctors and NGOs in the area. In Madanpur Khadar, a local NGO 'Sunshine' and 'Social Welfare Society' and *Pradhan* from the local village-level committee were identified for encouraging activities in the area. However, owing to technological and area specific challenges, involvement of these stakeholders also could not bring appreciable results in the activity. In Ambedkar Colony, the rollout of the model has shown appreciable results in a short span of on-field activity. This was due to the involvement of *Pradhan* of the village, who provided on-field assistance. This support essentially helped in strengthening engagement with the community and building up their confidence for the on-going activity in the area. Moreover, lessons from experiences at Madanpur Khadar helped in formulating strategy and initiating work at Ambedkar Colony as the work here started in succession of Madanpur Khadar.

12.2.3.2 Piloting Microfinance-Inclusive LEU Model in Bundelkhand

Learning from the lessons in Delhi, a modified version of the LEU model with Self-Help Groups (SHGs) and local NGOs was piloted in few remote villages of Chattarpur, and Hamirpur districts of Bundelkhand region in Central India. The financial barrier was overcome by engagement with the local women SHGs, who supported the community through microfinance loans for purchasing the filter. The technological barriers were addressed through selection of the location with low-TDS levels.

During the study period, appreciable results in filter adoption rates were obtained in rural Bundelkhand with a 13% penetration in 3 months (among 3216 no. of households), as compared to Delhi with only 1.6% success. As most of these filters need filter media replacement periodically, the role of supply chains is quite critical in supplying replenishments and not just one-time sales. Hence, the model can become unsustainable if the supply chains function erratically. Therefore, the filter manufacturers are expected to build closer tie-ups with supply chains and provide oversight on distribution channels for sustaining and expanding their market outreach among vulnerable populations.

The Model

To overcome the challenges of recruiting individuals as LEUs, the presence of a strong local NGO was considered as a critical factor for the selection of a location during the pilot rollout in Bundelkhand. Sumitra Samajik Kalyan Sansthan (SSKS) was selected to enhance the outreach of services to the community. Also, the confidence of the community in the local NGO would serve as a driver for behaviour change (Mishra 2014). Based on our experience during the Delhi roll out, behaviour change in the community is an essential component for wider dissemination of safe water services especially as these services are often not prioritised by the BoP. SSKS was also involved in selecting and providing technical and institutional support to the SHG.

The model is centered on the involvement of Jai Shankar Swayam Sahayata Samooh, a local woman Self-Help Group (SHG) in the role of the LEU. It was strategized that one entity (the identified SHG) would undertake a long-term business initiative and be the driving force in the identified geographical domain. It will leverage the efforts of local entrepreneurs, who can promote the filters in their available time to generate some extra income whenever and wherever possible (Fig. 12.2).

Tata Chemicals enabled the SHGs to procure the filters at a lower rate of Rs. 839, as the cost of the distributor was eliminated in this model. In one of the models, the filter was sold at the market price of Rs. 999 to the end customers, leaving the SHG with a profit of Rs. 160. An alternative model was implemented; wherein local shops were engaged with SHGs. Filters were also sold to these shopkeepers at a rate of Rs. 899. This resulted in an Rs. 60 profit to the SHG and an Rs. 100 profit to the local shopkeeper for each filter sold.

For widespread dissemination of safe water services among the community, in addition to upfront payment options, the model considers microfinancing options essential to overcome the financial barriers. Under microfinancing, the interested individuals from the target areas were provided with a loan to cover the initial installment of Rs. 500 from the SHG for booking of the filter. This amount was either leveraged through SHG funds or *Rashtriya Mahila Kosh* fund, which was reaching the SHGs through the active intermediation of SSKS. The remaining amount of Rs. 499 was given by the customer at the time of procurement of filter. The amount of Rs. 500 leveraged by SHG to the customer was repaid to SHG within 1–3 months. The SHG charged interest at the rate of 18% for the loan. The model follows similar lines as the delivery model for Delhi roll out for replacement of filter bulb made by

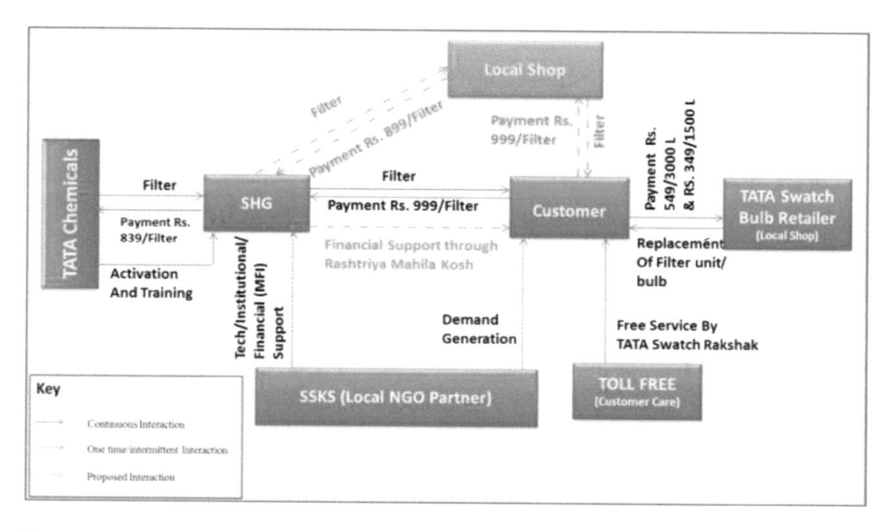

Fig. 12.2 Microfinanced LEU model schematic for Tata Swach

the local retailers and for activation and training services by Tata Chemicals. The project team facilitated the links between the various stakeholders.

On-the-Ground Activities

Demand creation is an essential component for wide-scale dissemination of services and long-term sustainability of the model. Prior to the rollout of Swach filter in the target locations, extensive efforts were put into creating demand for the product. Most of the times, the need for safe drinking water is not prioritised amongst the BoP population and, therefore, these needs remain unmet. There were regular awareness generation sessions conducted in Bundelkhand highlighting the importance of safe drinking water and how nano-based water filtration technology provides an affordable and effective solution to achieve the same. In addition, street plays were also organised at these locations. The team also used IEC material to facilitate information exchange with the community on importance and handling of safe drinking water. The community was also made aware of their drinking water quality by conducting on-field testing of water quality parameters including TDS using TDS meter and bacterial contamination using Aquacheck vials. The on-site testing of water quality had a massive impact on the communities. The approach for the field rollout was similar to the one adopted in Delhi.

The Challenge

Sustaining the supply chains remains a challenge. Tata Swach filters are not readily available in the nearby markets of Chhatarpur and Hamirpur. The nearest procurement of Swach for the on-going activity is made from the Ghaziabad region of Uttar Pradesh, which often consumes a lot of time in delivery of the filters. Time-consuming process of filter delivery to the target location is eventually leading to loss of interest in the community especially creating issues with the individuals who have paid for the services in advance. In order to speed up this process, and maintain the momentum created, DA has intervened and facilitated the process so that the local NGO is in close contact with manufacturer tracking the delivery of the orders. Efforts have also gone into making the manufacturer realise the potential of the market, which will also drive timely delivery of services to the growing client base in the study area.

12.3 Imperatives for Scaling up

The pilots lent themselves to a lot of learning on issues of access to water as well the support systems. The section below highlights some of the key lessons learnt from the experiences in Delhi and Bundelkhand.

- **Influencing behaviour change in the community for safe drinking water services**: Based on our understanding of the attitude of BoP towards safe drinking water services, this section of society is aware of the health impacts associated with contaminated water, however, the need for such services often gets neglected over other priorities. Therefore, it becomes imperative to influence the behaviour to seek safe drinking water. For this on the spot water quality testing, street play/awareness generation campaign can play a very vital role in bringing out changes at grass-root level, in creating much-needed awareness and behaviour changes amongst the public at large. It was observed that the best method to change people's mind set is on the spot water testing, wall paintings and street plays. In addition, sessions involving close interaction between the community and influential individuals from the medical field, water industry, etc., were helpful.
- **Engaging with local NGO/SHGs can better facilitate service delivery mechanisms**: Based on our experience from the on-field activities, it can be clearly concluded that the peer-influenced models are not very successful for providing safe drinking water services, however, similar models for other products such as of Amway has shown appreciable results in urban space. The reason for the same is because safe drinking habits are not often prioritised and hence the demand is very limited, which is usually perceived as a risk by the entrepreneurs as it questions the profitability of the business. Therefore, these entrepreneurs are only interested to work in models from where they can derive fixed salaries on the monthly basis. Engaging with local NGOs and SHGs who have been present in the community from a long time in similar kind of activities would be beneficial in terms of their understanding on the needs of the community and the confidence community has on them.
- On the **financial front**, the introduction of **microfinancing** with easy instalments-based schemes is a driver that accelerates uptake. Since the sale of filter is a consumer product, user preference and expectations from the product are keys to successful adoption. Building financial incentives along the supply chains is a key factor for sustaining and scaling up of the service delivery models.

- **Aesthetic aspects—to enhance product acceptability**: In addition to the filter parameters such as affordability, effectiveness in treating water, the rate of filtration and ease of filtration, change of **taste and aesthetics** catering to aspirations also are found to be critical parameters in accepting a treatment solution or a filter product.

12.4 Way Forward

The study highlighted the hypothesis that access to water is impeded by the lack of appropriate delivery systems to the Bottom of the Pyramid. The pilots clearly demonstrated that there is a demand for safe water. However, to convert this demand into changed behaviour and practice, there is a need for customers to see clear value in the product. This value creation can be perceived both in terms of the service offered and a perceptible difference from the raw water baseline as well as the financial value of the product, in terms of ease of payment, avoided health costs, etc.

In order to be able to service the BoP, it is important to develop and strengthen the delivery mechanism. The pilot showed that penetration and conversion were much higher in spaces where there was trust at the point of delivery. Building capacities of all points of the chain are important to ensure smooth delivery of the product and service.

References

Report

Vijaya Lakshmi, K., Nagrath, K., & Jha, A. (2011). *Access to safe water: Approaches for nanotechnology benefits to reach the bottom of the pyramid*. New Delhi: Development Alternatives Group.

Vijaya Lakshmi, K., Nagrath, K., & Mishra, S. (2013). *Access to safe water for the bottom of the pyramid*.

Online Document

DA. (2011). *Market potential analysis for water purifier using nanotechnology for the bottom of pyramid market*.

TARA. (2012). *Access to safe water for the bottom of pyramid: Strategies for disseminating technology research technology packaging study*.

WHO. (2011). *Estimated with data from Diarhhoea: Why children are still dying and what can be done*. The UNICEF fund.

Newsletter Article

Kumar, S. (2013, October). Innovative service delivery model for providing low cost safe drinking water solutions to rural India.
Mishra. S. (2014, March). Providing low cost safe drinking water solutions to rural Bundelkhand.

Chapter 13
Megaprojects as an Instrument of Urban Planning and Development: Example of Belgrade Waterfront

Slavka Zeković, Tamara Maričić and Miodrag Vujošević

13.1 Introduction

Megaprojects are considered as large-scale capital investments, single or multi-purpose. They include infrastructure projects, transport projects, economic development, and urban redevelopment (including waterfront redevelopment). This research is devoted to the analysis of UMPs as an instrument in the development of post-socialist cities, which are shaped by a mix of economic interests, socio-political and institutional framework.

Huge investments, large and diverse risks and impacts of megaprojects have led to a higher interest in their planning and management. The studies of megaprojects worldwide (del Cerro Santamaria 2013; Flyvbjerg et al. 2003; Kennedy 2013) show that the difference between the dominant regions of the Global North-West[1] and South-East is not as large as it appears. Megaprojects usually involve "exceptional" forms of governance, and do not go through the normal channels (Kennedy et al. 2014). The issues of high risks and uncertainty, cost underestimation and overruns, low public informing, lack of transparency, social and environmental impacts remain similar in countries with different institutional systems and level of economic development.

In the post-socialist countries, transitional changes have created new power relations between different groups involved in urban development and increased

[1] This is a more precise North–South division in terms of macro-regionalization.

S. Zeković (✉) · T. Maričić · M. Vujošević
Institute of Architecture and Urban & Spatial Planning of Serbia, Belgrade, Serbia
e-mail: slavka@iaus.ac.rs

T. Maričić
e-mail: tamara@iaus.ac.rs

M. Vujošević
e-mail: skomi03@gmail.com

S. Hostettler et al. (eds.), *Technologies for Development*,
https://doi.org/10.1007/978-3-319-91068-0_13

influence of private investors. This paper will analyze the conceptual framework of the UMPs in the context of post-socialist city. After presenting the theoretical background related to urban planning and governance of megaprojects and highlighting their role and common features, the research will focus on the experience of Belgrade Waterfront Project (BWP).

13.2 Theoretical Background

The key *research objectives* are the analyses of the theoretical background of the role of megaprojects as an urban development instrument, as well as empirical analysis of the BWP experience and its potential development impacts.

The theory and methodology of urban planning and megaprojects faces complex issue of social contextualization of urban planning and governance, investment, and regulations. Since the 1970s and 1980s, under the pressure of globalization, the cities have started economic transformation according to the "Post-Fordist" development of different services, properties, high technology, etc. Different nature of urban development is linked to the political goals (Scott and Storper 2015), which shape the urban theory of global/world cities (Sassen 2008; Cochrane 2006; Brenner 1998, 2004).

Castells (1972), Lefebvre (1970) and Harvey (1973) supported a concept of the city as a theater of class struggle, and citizens' rights to urban space. Brenner (1999), Cochrane (2006) and Harvey (2012) suggested the re-conceptualization of older concerns on urban politics and governance. Harvey (2012) pointed to neoliberal domination in the changing nature of political governance scales (from cities over states to the global level). In accordance to new political-economic doctrines, the reorganization on national level gave new inputs for urban governance. Harvey (1989) identified the shift from managerial into entrepreneurial governance, i.e., from the focus on urban services to the promotion of economic growth. These intentions are often realized through mega-development projects, speculative construction and political economy of place.

Top-down urban planning approach by megaprojects opens research on urban governance. Many researchers point to the top-down approach in planning UMPs and to the bottom-up public resistance to claiming urban spaces (Flyvbjerg et al. 2002; Davis and Dewey 2013; Kennedy et al. 2011). But, even with well-developed bottom-up methods, the management of complex systems is inherently problematic (Slaev 2017). Sellers (2002) criticized the central role of international business elites and the influence of external capital on urban-policy making. Swyngedouw (1996) indicates that the state seeks to attract capital through place-based interventions in urban regions. He argued that large urban development projects have "less democratic and more elite-driven priorities." Political impulses are very important in the creation of urban MPs, but top-down approach offers no possibilities for democratic negotiations.

As instruments of urban planning and development, megaprojects include high-technology, sophisticated and non-standard technology, contemporary design, and ICT management. MPs are initiated by global economic restructuring and policy-makers, and supported by neoliberal urban development policies, often with transnational financial support and top political structure. UMPs promote interests of various property developers, but usually with state mobilization of public funds. Gellert and Lynch (2003) claim that MPs require coordinated flows of international finance capital. The spread of gains for the society is connected with direct government commitment to public benefits (Fainstein 2009).

MPs include *substantive changes in the legal, economic and political framework*. They have a *special power and special status*, an inherent nature, and a "special regime" of implementation (Altshuler and Luberoff 2003). They are not always planned in advance, but integrated ex post into planning documents. MPs represent a mode of urbanization (Roy 2003) and a "collateral" instrument against illegal, irregular and informal construction in cities of the South.

A major problem in megaproject policy is misinformation about the costs and benefits, and high risks. The "*megaproject paradox*" includes risky scenarios, underestimated costs, overestimated benefits and revenues, undervalued environmental impacts, overvalued economic effects (Flyvbjerg et al. 2003), as well as legal and ethical issues. Flyvbjerg (2014b) argued about the "*iron law of megaprojects*"—exceeding the budget and time-frame, a lack of accountability, and delays. He also indicated on the so-called "*survival of the unfittest*", i.e., building the worst projects instead of the best.

The key challenges, risks, and uncertainties in planning of UMPs ought to be identified, considered, and managed, including complex nature of UMPs; external shocks; stakeholders; governance changes of contract conditions; new legal and financial instruments, etc.

13.2.1 Applied Approach

In analyzing the urban megaproject Belgrade Waterfront, we combined a contextually appropriate approach, benchmarking and some elements of the phronetic planning approach. These approaches focus on the syncretic forms of urban and development policies and the current discourse analysis in Serbia (Zeković et al. 2015a).

13.3 Example of the Belgrade Waterfront Project

As a result of a number of exogenous and endogenous factors, a collapse of strategic thinking, research, and governance exists in Serbia for almost three decades. In developmental terms, after the socio-economic growth and development in the 1980s, the dissolution of the former Yugoslavia, regional wars, international isolation and sanctions, Serbia's economy collapsed, with the uncertainty of recovery. Serbia is one of the most undeveloped European countries on the "inner peripheries of Europe" (Vujošević et al. 2010). Its integration into EU or "Europeanization" outside the Union depends on development perspectives, global trends and a new institutional context.

In such circumstances, large investments of an already over-indebted country into an expensive megaproject of the inner-city waterfront redevelopment could hardly be expected to receive wide expert and public support. However, the idea of the Belgrade Waterfront megaproject has already been announced to the public in 2012, in accordance with the "*fast-lane approach to investors*", with political statements that the "Tower Belgrade will become a new trade-mark of the capital city and Europe." Bancroft (2015) argues that the BWP embodies the promise of Belgrade's return to the world stage. After adopting the Agreement and Law on Cooperation between the Government of the Republic of Serbia and the Government of United Arab Emirates (2013) Serbian government founded the Belgrade Waterfront Company in 2014 to mobilize public funds for the BWP implementation.

The BWP was integrated ex-post into the master plan of Belgrade in 2014, and Belgrade Waterfront Spatial Plan was adopted in 2015. The BWP is a Dubai-inspired project of the old city's waterfront redevelopment with little public resistance. However, the Academy of Serbian architects adopted a Declaration on the BWP in 2015, with arguments against the project.

In accordance with the specific ordinance (2006, 2012), the BWP is verified as a national priority, which illustrates a dominant model of public–private partnership with a national/metropolitan influence, despite the influence of an international private investor. The main legal precondition for the realization of the BWP was the adoption of a *lex specialis*—a Law on establishing the public interest and the special procedures of expropriation and issuance of construction permits (only) for the BWP (2015). Another legal precondition was a Joint Venture Agreement (JVA) between the Republic of Serbia, Belgrade Waterfront Company and investors from the UAE, without a tender process. The Ministry of Building, Traffic and Infrastructure has issued construction permits for the first two towers and their construction has already started.

The main goal of the BWP is to activate the waterfront and develop a modern urban center, thus promoting an international image of the Belgrade. *The key objective* of the Belgrade Waterfront Spatial Plan is to transform a neglected area into a modern city center. *The general objectives* are: protection of river Sava biodiversity, landscape improvement; revitalization of cultural heritage; better life quality; affirmation of Belgrade such a tourist destination; modern commercial offer; construction of major transport systems and infrastructure.

The BWP envisages the construction of two million m^2 on 177.27 ha: 6128 flats (one million m^2), commercial spaces (main tower 210 m high, shopping mall, several tall buildings), social, cultural, recreational and free spaces. There are predictions about cleaning the riverfront, old buildings, railway infrastructure, abandoned ships, environmental clean-up, etc. Total investment in BWP is 3 billion EUR for three phases (8–30 years).

The decision-making regarding megaprojects in Serbia is exclusive and elite driven. This is visible: (1) by innovations in the existing Planning and Construction Act (PCA), (2) by appreciation of the new *lex specialis* (expropriation for private elite hi-tech housing and commercial purposes for the BWP), (3) by verification of the BWP as a national strategic priority, (4) by property development regulation, (5) local communities are excluded from decision-making and poorly informed about BWP. Political leaders and the mayor of Belgrade provide the majority of the information regarding the BWP. Dogan (2015) argued that a strong national initiative in this project represents a top-down approach to the regeneration of the wider Savamala district. Eror (2015) argued that "it's a state-driven model" called "top-down" or "hyper" gentrification.

13.3.1 Benchmark of Development Impacts of the BWP

The absence of accountability is evident in the BWP decision-making. Monitoring and control systems are insufficient, as well as the approaches for evaluating main social, economic and environmental impacts.

Benchmarking of possible development impacts of the BWP include intensive social impacts (raising social inequalities, gentrification, involuntary resettlement, networking of the key actors), intensive impacts on national level (overuse of public funds, limiting the State in making laws that are incompatible with the BWP interests), high public financial risk, strong urban transformations, intensive demographic growth (17,700 new inhabitants), low development and economic effects, low transparency and public participation, environmental impacts, and others (Table 13.1). Policy-makers promoted the BWP, emphasizing its role in creating employment (13,169–200,000 new employees), promoting tourism, using domestic inputs, improving productivity, growth competitiveness, high-quality services, etc.

Due to the complexity of the BWP, it would be difficult to provide proofs for such possible impacts and their exact scopes. The lack of a feasibility study for the BWP and non-transparent data are reasons for the insufficiently reliable and quantified assessment of its effects.

Total investment in capital infrastructure and utilities is 400 million EUR. The cost of activating the location is between 790 and 1000 million EUR (Kontrapress 2014), while expropriation costs will reach 7.1 million EUR. The Belgrade Waterfront Spatial Plan states that the revenues from urban development land and participation for infrastructure are 1.03–1.33 billion EUR, while the potential revenue from urban land (land value) varies from 168 to 336 million EUR.[2]

Benchmarking of regulation indicates some open *regulatory issues* which may have potential impacts on the development effects of the BWP, such as: (1) The

[2]The expected project benefits do not include property tax, capital gain tax, real estate transfer tax, levies, income tax, fee for use of public goods and others.

Table 13.1 A preliminary impact assessment of the BWP

Development impacts	High intensity	Medium intensity	Low intensity	Level	Impact evaluation
Development effects			X	National, local	–
Transparency and public participation			X	Local	–
Raising social and spatial inequalities	X			Local	–
Gentrification	X			Local	–
Demographic impacts	X			Local	–
Displacement effects	X			Local	–
Environmental impacts		X		Local	+ and –
Economic effects			X	National, local	+ and –
Government independence in law-making	X			National	–
Public financial risk	X			National	–
Urban and spatial transformations	X			Local	+ and –
Technological modernization		X		Local	+
Regulatory regime impacts	X			National	–

Decision of Serbia's Commission for State Aid as a moot point. The Stabilization and Association Agreement (2008) and EU regulations *prohibit the disturbance of competition by state aid for the elite-housing and commercial spaces; (2) No pre-feasibility study, no scientific analysis, and no urban study*; (3) A discrepancy in choice of institutions included in the evaluation of urban construction land, (4) Unclear dynamics of investment, total sum, and agreed (disproportionate) share of the strategic partner in financing the buildings in the BWP, and a real risk from the lack of finances; (5) The regulatory regime implicates public risks and costs, an inverted order in the preparation of the pre-feasibility study and different appraisals, implementation postponement, and (6) Discrepancies in the Law on accepting the agreement between the governments of the Republic of Serbia and United Arab Emirates with the Constitution (e.g., agreements, programs and projects according to this law are not subject to public tender, while the Constitution prescribes that international agreements must be in accordance with the market economy).

The introduction of specific legal and policy instruments in Serbia under neoliberal economic pressures is a key source of the future change in the metropolitan tissue by the BWP. Potentially negative development impacts of the BWP in some legal aspects of the urban construction land might comprise: (1) Leaseholds of public urban land without fees to the Belgrade Waterfront Company which is in dominantly private ownership of BWP investor (by Regulation, 2012); (2) Right of lease to the Belgrade Waterfront Company *gratis* will be converted into the right of ownership. This Company can transfer the right of ownership to other parties without a fee (after constructing building/s and obtaining a use-permit) in accordance with *lex specialis*, Regulation (2012); (3) The obligation of Serbia and the City of Belgrade to finance and build all external and internal capital infrastructure till December 31, 2019; (4) An enactment of necessary legislation allowing full set-off of all land development fees against public land development costs on the project level. Land development fees will be governed by a separate agreement between the City of Belgrade and national institutions. If Serbia does not fulfill contractual obligations, it has to pay damages to the strategic partner; (5) The final calculation of the costs between investors and the City of Belgrade will be made after completing the construction of all planned facilities, but without a stated period; (6) Serbia could not change the BWP plan without the approval of a strategic partner, while the partner can change some parts of the plan; and (7) Serbia has an obligation to adopt the necessary changes to other laws that are desirable according to the JVA, and it can limit the independence of the national government in passing the laws.

Citizens are mainly excluded from the decision-making, including low level of public informing. The protests of citizens and NGOs reflect insufficient transparency and democracy in the planning of BWP.

13.4 Recommendations for Future Research and Application

In many regions of the South-East, state policies support urban development compatible with elite tastes and consumption that promotes socioeconomic inequalities, thus enabling global finance capital to shape the city (Watson 2012). This "privatization" of planning, as Shatkin (2011) calls it, through megaprojects tends to undermine the public administration of urban space and replace local authority with private governance. Recent research of megaprojects in cities of the South (Kennedy et al. 2014) showed the decreased significance of local government within the process of economic development.

The empirical findings of the BWP open possibility for a new insight into planning and appraisal of UMPs, that includes an alternative/improved approach and the following recommendations: more transparency and real public participation, improved

Table 13.2 Preliminary assessment of differences between traditional and "alternative"/improved approach to planning UMPs

Type of tools/instruments	Alternative approach	Traditional/conventional approach
Transparency and public participation in decision-making	Increased, with bottom-up approach	Mostly minor, with top-down approach
Performance specification	Goal-driven approach	Technical solution-based/driven approach
Better regulatory framework	Improvement of regulatory framework	Inversion between feasibility study and choice of regulatory regime (e.g., lex specialis decreases the role of feasibility study)
Pre-feasibility study	Required, independent peer-review	Required, independent peer-review is rarely done
Risk analysis	Inclusion of risks, acceptable level of public risks	Ignores risks, unacceptable level of public risks

regulatory framework, less use of private international risk capital, compilation of pre-feasibility study and appraisals, public bid for possible involvement of the private sector in financing the capital city infrastructure, and limitation of the state guarantee to lenders for funding the UMPs, especially in joint projects. The improved approach to planning UMPs includes a goal-driven approach in the preparation of the feasibility study and decision-making instead of traditional technical-driven approach (Table 13.2). This improved approach is closer to the general context in the countries of the North-West and a key instrument in their planning systems. The improved approach to planning UMPs should include

(a) Improvement of the urban planning system, better evaluation methods, planning evaluation of alternatives, implementation policy, and more innovative urban land policy tools (Zeković et al. 2015b);
(b) Establishment of minimum international standards for the change of national legislation of the UMPs;
(c) More transparency in decision-making, with the real and wide participation of different stakeholders, their involvement in "policy re-design" and formulation and evaluation of alternatives. Outdated and often inappropriate in the context of changing urban environments, bureaucratic approaches are still predominantly used in the major part of the Global South (UN-Habitat 2009);
(d) Performance of UMPs implies *a goal-driven approach* in the preparation of the feasibility study which differs from the traditional *technically driven and top-down approach* in decision-making. In the planning of capital urban projects, significant elements are democratic legitimacy (both top-down and bottom-up simultaneously), technical and economic rationality, social and environmental

acceptability. The performance of UMPs should derive from key objectives of policies/plans and public interest. The traditional *technical solution-driven approach* in decision-making of MPs characterizes the Global South-East, but sometimes also appears in the Global North-West;

(e) Setting better regulatory framework involves elimination of policy risks and/or their inclusion before decision-making (e.g., risk assessment), underestimating costs and overestimating benefits, as well as different types of ex ante impact assessments for evaluating, mitigating and balancing different impacts.

Generally, tools suggested in the “alternative”/improved approach are mainly used in the North-West, while traditional instruments predominate in the South-East. The new approach in planning, governing and implementing UMPs also requires multidisciplinary approach, critical analysis of the conventional approach, introduction of measures for improved policies and planning, and determination of the interplay between the different pools of power.

13.5 Conclusions

Our analysis highlights the differences in institutional, social and economic environment that shape the Belgrade Waterfront Project. At the same time, the BWP induced a substantial change of institutional framework (introduction of specific legal and policy instruments) under neoliberal economic pressure, which represents a key source of future changes in the metropolitan tissue. The benchmarking of the developmental effects of the BWP especially underlined: intensive social inequalities, marginal social mobilization of the key actors and stakeholders, intensive impacts on national level, high displacement effects, high public financial risks, strong urban transformations, low development and economic effects, low transparency and public participation, environmental impacts, etc.

The specific nature of the BWP requires specific instruments, including legal, financial, economic, construction, environmental, and more innovative and flexible urban land instruments. The proposed recommendations for their improvement would result in better development effects for the city.

Decentralization of regulation powers is important from the standpoint of planning, decision-making, governance, control, and implementation of megaprojects. It provides recommendations for future research and application, for a continuing in-depth analysis to mitigate all consequences of the UMPs, including determination of the interplay between different pools of power (“from power to tower”).

Acknowledgements This paper is the result of research carried out within the scientific project *Support to Process of Urban Development in Serbia* (SPUDS) funded by the SCOPES program of the Swiss National Science Foundation, (2015–2018) and the scientific project *The role and implementation of the national spatial plan and regional development documents in renewal of strategic research, thinking and governance in* Serbia No. 47014 financed by the Ministry of Education, Science and Technological Development of Serbia.

References

Altshuler, A., & Luberoff, D. (2003). *Mega-projects: The changing, politics of urban public investment*. Washington, D.C., Cambridge, Mass: Brookings Institution Press.

Bancroft, I. (2015). Op-ed: Beograd na Vodi, the White City's latest rupture. *Balkanist Magazine*. 27, November 2015. http://balkanist.net/beograd-na-vodi-the-white-citys-latest-rupture-2/. Accessed November 25, 2015.

Brenner, N. (1998). Global cities formation, and state territorial restructuring in contemporary Europe. *Review of International Political Economy, 5*(1), 1–37.

Brenner, N. (1999). Globalisation as reterritorialisation: The re-scaling of urban governance in the European Union. *Urban Studies, 36,* 431–451.

Brenner, N. (2004). *New state spaces: Urban governance and rescaling of statehood*. Oxford, New York: Oxford University Press.

Castells, M. (1972). *La Question Urbaine*. Paris: Maspero.

Cochrane, A. (2006). *Understanding urban policy: A critical introduction*. Oxford: Blackwell.

Davis, D. E., & Dewey, O. F. (2013). How to defeat an urban megaproject: Lessons from Mexico City's airport controversy. In G. del Cerro Santamaria (Ed.), *Urban megaprojects: A worldwide view,* research in urban sociology (Vol. 13, pp. 287–315). Bingley: Emerald books.

del Cerro Santamaria, G. (Ed.). (2013). Urban megaprojects: A worldwide view. *Research in Urban Sociology, 13* (Bingley: Emerald books).

Dewey, O. F., & Davis, D. (2013). Planning, politics, and urban mega-projects in developmental context: Lessons from Mexico City's airport controversy. *Journal of Urban Affairs, 35*(5), 531–551.

Dogan, E. (2015). City on the rise: Mega projects vs. public resistance. In A. Gospodini. (Ed.), *Proceedings of the International Conference on Changing Cities II Spatial, Design, Landscape & Socio-economic Dimensions.* University of Thessaly, Porto Heli, Greece, June 22–26, pp. 743–750.

Eror, A. (2015, December 11). Belgrade's "top-down" gentrification is far worse than any cereal café. *The Guardian.* http://www.theguardian.com/cities/2015/dec/10/belgrade-top-down-gentrification-worse-than-cereal-cafe. Accessed December 21, 2015.

Fainstein, S. (2009). Mega-projects in New York, London and Amsterdam. *International Journal of Urban and Regional Research, 32*(4), 768.

Flyvbjerg, B. (2009). Survival of the unfittest: Why the worst infrastructure gets build and what can do about it. *Oxford Review Economic Policy, 25*(3), 344–367.

Flyvbjerg, B. (2013). How planners deal with uncomfortable knowledge: The dubious ethics of the American Planning Association (APA). *Cities, 32,* 157–163.

Flyvbjerg, B. (2014a). *Megaproject planning and management: Essential readings* (Vol. 1–2). MA, USA: Edward Elgar Publishing.

Flyvbjerg, B. (2014b). What you should know about megaprojects and why: An overview. *Project Management Journal, 45*(2), 6–19.

Flyvbjerg, B., Bruzelius, N., & Rothengatter, W. (2003). *Megaprojects and risks: An anatomy of ambition*. Cambridge: Cambridge University Press.

Flyvbjerg, B., Holm, M. K. S., & Buhl, S. L. (2002). Underestimating costs in public works projects: Error or lie? *Journal of the American Planning Association, 68*(3), 279–295.

Gellert, P., & Lynch, D. B. (2003). Megaprojects as displacements. *International Social Science Journal, 175,* 15–25.

Harvey, D. (1973). *Social justice and the city*. London: Edward Arnold.

Harvey, D. (1989). From managerialism to entrepreneurialism: The transformation in urban governance in late capitalism. Geografiska Annaler. *Series B Human Geography, 71*(1), 3–17.

Harvey, D. (2012). *Rebel cities: From the Right to the city to the urban revolution*. London: Verso.

Kennedy, L. (2013). Large scale projects shaping urban futures. A preliminary report on strategies, governance and outcomes based on eight case studies in four countries. Bonn: EADI.

Kennedy, L., Robbins, G., Bon, B., Takano, G, Varrel, A., & Anrade, J. (2014). *Megaprojects and urban development in cities of the south. Thematic Report, 5* (Chance2Sustain).
Kennedy, L., Robbins, G., Scott, D., Sutherland, C., Denis, E., & Anrade, J. (2011). The politics of large-scale economic and infrastructure projects in fast-growing cities of the south. *Literature Review, 3* (Chance2Sustain).
Kontrapress. (2014, October 30). Ispod površine Beograda na vodi: Planovi i posledice (Underground of Belgrade Waterfront: Plans and consequences), http://www.kontrapress.com/clanak.php?url=Ispod-povrsine-Beograda-na-vodi-Planovi-i-posledice. Accessed November 18, 2015.
Lefebvre, H. (1970). *La Révolution Urbaine*. Paris: Gallimard.
McFarlane, C. (2010). The comparative city: Knowledge, learning, urbanism. *International Journal of Urban and Regional Research, 34,* 725–742.
McFarlane, C. (2011). Assemblage and Critical urban Praxis. *City, 15,* 204–224.
Rankin, K. N. (2011). Assemblage and the politics of thick description. *City, 15,* 563–569.
Roy, A. (2003). Urban informality. Towards an epistemology of planning. *Journal of the American Planning Association, 71,* 147–158.
Sassen, S. (2008). *Territory, authority, rights; from medieval to global assemblages*. Princeton: Princeton University Press.
Scott, A. J., & Storper, M. (2015). The nature of cities: The scope and limits of urban theory. *International Journal of Urban and Regional Research, 39*(1), 1–15.
Sellers, J. M. (2002). *Governing from below: Urban regions and the global economy*. Cambridge, New York: Cambridge University Press.
Shatkin, G. (2011). Planning privatopolis: representation and contestation in the development of urban integrated mega-projects. In A. Roy & A. Ong (Eds.), *Worlding cities: Asian experiments and the art of being global* (pp. 77–97). Chichester: Wiley-Blackwell.
Slaev, A. D. (2017). The relationship between planning and the market from the perspective of property rights theory: A transaction cost analysis. *Planning Theory, 16*(3), 404–424.
Swyngedouw, E. (1996). Reconstructing citizenship, the rescaling of the state and the new authoritarianism: Closing the Belgian mines. *Urban Studies, 33*(8), 1499–1521.
UN-Habitat. (2009). *Planning sustainable cities. Global report on human settlements*. London: Earthscan.
Vujošević, M., Zeković, S., & Maričić, T. (2010). *Postsocijalistička tranzicija u Srbiji i teritorijalni kapital Srbije. Stanje, neki budući izgledi i predvidljivi scenariji (Postsocialist transition in Serbia and territorial capital of Serbia. Condition, some future prospects and foreseeable scenarios).* Beograd: Institut za arhitekturu i urbanizam Srbije.
Watson, V. (2012). Seeing from the South: Refocusing urban planning on the globe's central urban issues. *Urban Studies, 46*(11), 2259–2275.
Zeković, S., Vujošević, M., & Maričić, T. (2015a). Spatial regularization, planning instruments and urban land market in a post-socialist society: The case of Belgrade. *Habitat International, 48,* 65–78.
Zeković, S., Vujošević, M., Bolay, J. C., Cvetinović, M., Živanović-Miljković, J., & Maričić, T. (2015b). Planning and land policy tools for limiting urban sprawl: Example of Belgrade. *Spatium, 33,* 69–75.

Legal Acts and Plans

Lex specialis—The Law Establishing the Public Interest and Special Procedures of Expropriation and the Issuance of Construction Permit for the Project "Belgrade Waterfront". Official Gazette of RS, No. 34/2015.
Ordnance of pre-feasibility study. Official Gazette of RS, No. 1/2012.

Ordnance/Pravilnik o metodologiji i proceduri realizacije projekata od značaja za Republiku Srbiju. Official Gazette of RS, No. 59/2006 and 1/2012.

Regulation/Uredba o uslovima i postupku otuđenja ili davanja u zakup građevinskog zemljišta u javnoj svojini RS po ceni manjoj od tržišne ili bez naknade. Official Gazette of RS, No. 67/2011, 85/2011, 23/2012 and 55/2012.

The Act confirming the Agreement on Cooperation between the Government of the Republic of Serbia and the Government of United Arab Emirates (Zakon o potvrđivanju Sporazuma o saradnji između Vlade Republike Srbije i Vlade Ujedinjenih Arapskih Emirata). Official Gazette of RS, No. 3/2013. http://www.salter.rs/zakoni/me%C4%91unarodni/zakon-o-potvr%C4%91ivanju-sporazuma-o-saradnji-izme%C4%91u-vlade-republike-srbije-i-vlade. Accessed November 13, 2015.

The Agreement on Cooperation between the Government of the Republic of Serbia and the Government of United Arab Emirates. (2013). Official Gazette of RS, No. 3/2013.

The Amendments on the Master plan of Belgrade (Izmene i dopune Generalnog plana Beograda 2021). (2014). Official Gazette of the City of Belgrade, No. 70/2014.

The Belgrade waterfront Spatial plan (Prostorni plan područja posebne namene uređenja dela priobalja grada Beograda – područje priobalja reke Save za projekat "Beograd na vodi"). Official Gazette of RS, No. 7/2015.

Chapter 14
What Can the South Learn from the North Regarding the Implementation of IoT Solutions in Cities? The Case of Seoul-Born Smart Transportation Card Implementation in Bogota

Maxime Audouin and Matthias Finger

14.1 Introduction

Over the past decade, the urban population of the Global South has been growing at an average rate of 1.2 million people per week (UN Habitat 2013), bringing developing countries to center stage on issues relating to urban development. Due to continuous urbanization and growth, cities have become increasingly complex environments (McHale and al. 2015; Hillier 2009), calling for the development of innovative solutions to ensure service delivery and well-being of their citizens. Collecting and processing the data produced by the operation of urban infrastructure systems, such as transportation, telecommunication, water, waste, or energy, can actually be seen as one of those solutions. Creating an interconnected network (Internet) of physical components from urban infrastructure systems (Things), that is to say the Internet of Things (IoT) (Dijkman and al. 2015), is indeed viewed as an efficient way to deal with much of the complexity associated with urban systems (Zanella et al. 2014), and for enhancing development. As a matter of fact, IoT solutions appear as having the potential to support the development of solutions contributing to make cities more resilient, sustainable, safe, and inclusive (SDG 11) and more generally to environmental sustainability (MDG 7).

However, most of IoT solutions usually originate from developed countries, part of what is often referred to as the "Global-North", and one might wonder their degree of transferability to the "Global-South". This chapter thus seeks to explore the conditions required to reproduce IoT "success stories" from the North to the South by answering the following question: **what can the South learn from the North regarding the implementation of successful Internet of Things (IoT) solutions?**

M. Audouin (✉) · M. Finger
Chair Management of Network Industries (MIR), EPFL, Lausanne, Switzerland
e-mail: maxime.audouin@epfl.ch

M. Finger
e-mail: matthias.finger@epfl.ch

S. Hostettler et al. (eds.), *Technologies for Development*,
https://doi.org/10.1007/978-3-319-91068-0_14

To answer this question, we present the cases of development of the same IoT solution, being a smart transportation card, in Seoul (North) and Bogota (South), based on archival data and data collected from semi-structured interviews with stakeholders involved in the projects. We then conduct a comparative analysis of the two cases using a conceptual framework building on co-evolution between technology and institutions theory, to understand the extent to which technological innovation can be a driver for institutional changes in urban transportation systems, and vice versa.

14.2 Seoul Case

Seoul, the economic and political capital of the Republic of Korea, is one of the most populated urban areas in the world. In 1960, the Greater Seoul metropolitan area was home to 5.2 million of inhabitants (Bae and Richardson 2011); it is estimated that nowadays, about 23 million people live in Seoul Metropolitan Area, accounting for approximately half of the South-Korean population (Cervero and Kang 2011).

Consequent transportation problems resulted from the exponential growth Seoul experienced. Driven by strong industrial development, Seoul's demographic growth was followed by an impressive economic growth. The per capita income (in 2004 US Dollars), that was about US$ 311 in 1970, rose to US$ 7378 in 1990, and finished 2002 at US$ 12,531 (Allen 2013). Automotive manufacturing became an especially powerful sector (Samsung Motors, Hyundai, Kia Motors), and because of the rising household economic power, more and more citizens started to buy private cars, eventually leading vehicle ownership to rise from 2 cars per 1000 persons in 1970 to 215 per 1000 persons in 2003 (Pucher et al. 2005). Unfortunately, Seoul's road infrastructure did not adapt to this impressive increase in car ownership. As a result, congestion became significant, leading to lower travel speeds and increased travel times for all road vehicles, ultimately becoming synonymous of decreased efficiency of the bus system, on which Seoul was heavily dependent at that time.

To tackle the congestion problem, Seoul Metropolitan Government (SMG) undertook the construction of an urban rail system, which was finally inaugurated in 1980. The new Seoul metro system constituted an alternative mode of transportation for Seoul citizens, who graciously welcomed its development. Unfortunately, this new solution did not act in favor of the Seoul bus system (Pucher et al. 2005). Indeed, due to the convenience and reliability of the new metro system, as well as the affordability of private cars, bus quickly became the less preferred means of transportation in Seoul, leading to a rapid decline in bus usage. Because of the decreasing number of passengers on buses having shifted to rail or private cars, bus companies operating under a license from SMG, also motivated by a lack of control from Seoul's authorities, started to do whatever was most profitable for them. Hence, it was not unusual at that time to see bus-operating companies cancel their non-lucrative routes without notice, or bus drivers drive recklessly to put as many passengers as possible onboard in order to generate more revenues, flouting the safety and comfort of their passengers (Allen 2013). As a direct consequence, 31 bus companies went bankrupt from 1995 to 2002 (Pucher et al. 2005).

Due to the urban sprawl of the Korean capital, resulting from its exponential growth after the post-Korean War, most of Seoul urban dwellers experienced long commutes, made of a combination of multiple trips, being synonymous of multiple transfers between different transport modes available, or within the same transport mode but between different operators. Because of the inexistence of an integrated fare system at that time, public transport users in Seoul were spending tremendous amounts of money on public transportation fare payments. Indeed, each time a user had to transfer, he had to pay an additional fare. Evidently, there was an urgent need for fare integration across the whole metropolitan area, and across all bus and metro operators. Consequently, this became one of the main campaign promises of Mr. Lee Myung-Bak, who was elected as Mayor of Seoul in the 2002 municipal elections. To support this integration, a new fare scheme was devised by SMG, as part of a broader public transportation reform, which would be supported by an integrated smart card ticket system (Park and Kim 2013). The first step of the creation of an integrated transport system was to change the ownership of the bus operation system in order to create a semi-public bus system.

To do so, SMG first created the Public Transport Promotion Task Force, led by the head of Seoul's transportation sector, which was composed of Seoul city officials and researchers from the Seoul Development Institute, who were to conduct the transportation reform from an expert point of view. Following the Institute's urban transportation research division head guidelines and based on the advice from the Public Transport Promotion Task Force, the Seoul Bureau of Transportation and the Transportation Policy Advisory Committee then created the Bus System Reform Citizen Committee (Kim et al. 2011), composed of 20 professionals from public authorities, industry and civil society, which aimed at solving conflicts between the different involved stakeholders (Kim and Dickey 2006). After a long series of formal and informal meetings between the CEOs of bus companies and SMG, the newly elected Mayor Lee Myung-Bak and the Chief of the Seoul Bus Transport Association, representing the 57 remaining bus-operating companies in Seoul, signed an agreement approving the bus reform and stating the change of the operational content of the companies' operating licenses. Once signed, it was just a matter of time to have the IoT technological vector (smart card) implemented, that would support the creation of a new integrated distance-based Automatic Fare Collection (AFC) system, allowing users to freely transfer across and between modes (Kim and Shon 2011).

The device, named T-money, was implemented and operated by a Special Purpose Company, called Korea Smart Card Corporation (KSCC) led by LG CNS, a subsidiary of the LG Group, and finally inaugurated on July 1, 2004 (Audouin et al. 2015). As part of the Public–Private Partnership (PPP) concluded between SMG and KSCC, the whole investment was performed by the private sector. The card itself is a plastic card embedded with a Central Processing Unit (CPU) that can store and transmit data when in contact with dedicated card readers, using Radio-Frequency Identification (RFID) technology (Blythe 2004). When implemented, the T-money card was a prepaid card, so users needed to top-up cash on their card at dedicated machines to be able to travel. Users would need to tap-in their card on card readers when entering a bus or metro station, and tap-out when exiting, so the AFC system would calculate the total distance traveled and deduce the corresponding fare from their T-money balance. Ultimately, the T-money sought to streamline human traffic flows at metro gates, to reduce bus delays caused by cash transactions for fare payments at vehicle gates, and as an IoT device, to ensure complete transparency of bus drivers operations, and obtain data about public transportation use by the citizens (Park and Kim 2013). The card was widely embraced by the population, as in 2013 it is estimated that it was used for 97.1% of the 32 million trips processed daily on Seoul public transportation system (SMG 2014).

14.3 Bogota Case

With almost 10 million inhabitants, Colombia's capital is the most populated city in the country (Munoz-Raskin 2010). The economic situation in Colombia, specifically in Bogotá, is significantly less developed than Korea and can be characterized as belonging to the Global South (Cervero 2013). Prior to 2000, the Metropolitan Area of Bogotá was characterized by a fairly complex transportation system. Bus services were provided by 64 different companies operating around 21,000 vehicles citywide. Private bus operators were often leasing their routes (previously obtained on concessions-based contracts) to third-party bus owners, creating a complex structure for bus operations (Cain et al. 2006). At that time, most of the bus-operating companies used obsolete vehicles, and showed very little concern for public safety. Bus stops were actually depending on the willingness of the bus drivers to stop, resulting in an anarchic bus system. As the income of bus drivers was directly linked to the number of passengers they carried, Bogota's bus system was synonymous of chaotic competition between operating companies, also known as the "penny war" (Ramos 2015), resulting in serious safety issues for pedestrians and other road users (World Bank 2010).

To address the poor condition of public transport in Bogota, the 1998 newly elected Mayor, Mr. Enrique Peñalosa, proposed a Bus Rapid Transit (BRT) system, called TransMilenio (Ardila 2004), in cohesion with the Bogota Mass Transport System plan of the Colombian National Development plan (Lara and Gutierrez 2012). TransMilenio was a citywide, city-owned system that was supposed to offer speed and convenience to its users. Buses were to run in dedicated corridors to avoid traffic flows, and riders were to purchase their tickets at the entrance of bus stations instead of upon entering buses. The project started in 1998 and was inaugurated in December 2000. Additional corridors opened every year until 2006 (Heres et al. 2013). Old buses were progressively taken out of these main corridors ensuring a smooth transition so that eventually TransMilenio was the only public transportation remaining on the main routes. Through concessions, TransMilenio outsourced the operation of the BRT and feeder line buses. Tendering processes were implemented where pre-TransMilenio era companies could bid if they had previously demonstrated experience operating the city transportation system. This process motivated the birth of new entities, composed of former bus companies partnering with firms from other industries responsible for cash investments, and high competition between them.

Because it improved the efficiency of the overall transportation system, reduced travel times and afforded significant cost saving for citizens (Bocarejo et al. 2014), TransMilenio was first considered a success for the city of Bogota. However, Bogota's public authorities failed in using the development of TransMilenio to create a city-wide integrated transportation system. While TransMilenio ran in a regulated manner on dedicated bus lines, pre-Transmilenio collective public transport, still completely unregulated, actually continued running in parallel on streets (Ramos 2015), creating competition between the BRT and pre-TransMilenio collective transport. Public authorities did not also use TransMilenio to create a citywide integrated fare system. Indeed, a dedicated fare system, the blue and red smart cards operated by Angelcom (El Espectador 2014), was specifically created for TransMilenio, that did not allow users to transfer freely to and from non-TransMilenio buses. To address the cruel lack of integration that was characterizing Bogotà public transport system at that time, the city of Bogotá proposed in 2009, as part of the Bogotá Mobility Master Plan, the development of the SITP (Integrated Public Transport System of Bogota). At the core of the SITP lied the development of an integrated fare system, which was to be supported by a unique payment method, aimed at improving local traveling conditions (SDP 2009). In July 2011, the city of Bogotá awarded Recaudo Bogotá, a concession composed of three different shareholders, namely Citymovil (60%), Land Developer (20%), and LG CNS (20%), for the operation of the fare collection and user information of SITP. LG CNS was in charge of providing an automatic fare collection system (AFC), aiming at integrating the TransMilenio routes with the new feeder routes, that were going to be developed later to replace the old unregulated feeder bus lines, dating from the pre-TransMilenio era.

The funding of Recaudo Bogotá was secured through a $176 million financing package loan by the World Bank, HSBC and Korean Banks (Shinhan and Woori Bank). The loan aimed at enabling Recaudo Bogotá to develop and introduce an easy-to-use electronic-payment mechanism supported by a smart card named "Tu Llave" card (standing for "Your key" in Spanish), which would eventually eliminate cash from the system, and enable free transfer between SITP bus routes, resulting ultimately in increased efficiency of the system, and security for users (IFC 2012). From a technological point of view the "Tu Llave" card works exactly as the T-money card as they both use Infineon's Security Microcontroller SLE66 with 4K Memory.

As part of the SITP, an important part of the city bus system was also redesigned, forcing most of the pre-TransMilenio bus companies operating in parallel to the BRT to stop operations, and to conform to the same competitive concession arrangements and regulations devised under TransMilenio (World Bank 2014). Those newly regulated bus-operating companies, who for many started to operate as TransMilenio feeder lines, combined with TransMilenio, formed the SITP network, on which the only mode of payment accepted was the "Tu Llave" card. As part of SITP, bus drivers also got a fixed salary, independent of the number of passengers transported, announcing the end of the "penny war" on the SITP network.

However, the implementation of the "Tu Llave" card did not happen as smoothly as planned. Indeed Recaudo Bogota and Angelcom, for which the blue and red smart cards were still in place on the TransMilenio network, had not reached any agreement. Hence, while the "Tu Llave" card was accepted on all newly regulated bus routes, it remained unaccepted at most TransMilenio bus stations. Recaudo and Angelcom finally reached an agreement in 2013 where Recaudo was to pay for the upgrade of all the TransMilenio card readers to make them "Tu Llave" compatible (El Espectador 2014). Since September 2015, both cards provided by Recaudo and Angelcom are accepted on the whole TransMilenio network. "Tu Llave" card users are thus able to access SITP buses (including TransMilenio) and Bogotá Metro, as well as to freely transfer between those. Currently, the "Tu Llave" card is used (and has to be used, as it is the only mean of payment) in all SITP buses and BRT system. The average numbers of daily transactions, including validations, recharge and transfer, processed by the "Tu Llave" card is around 7 million. In 2015, during the Korea–Colombia business forum, the Colombian President, Juan Manuel Santos, recognized the significant role LG CNS played in the development of the Bogotá transportation system (Business Korea 2015). However, it was estimated at the time this paper was written that approximately 2500 buses in Bogota were still operating as unregulated, thus outside of the SITP, and that Bogota's City Hall was continuously fighting to regulate them, to ultimately have the SITP cover the entire Bogota public transport system.

14.4 Analysis

Because these two cases deal with the introduction of the same technological innovation (smart transportation card) by the same technological actor (LG CNS), they can provide a good basis for comparative analysis. The following analysis is based on the alignment (sometimes referred to as coherence) framework from the co-evolution between institutions and technology literature (Finger et al. 2015). The framework acknowledges that infrastructure systems are "co-evolving" as a result of evolution in technologies on the one hand and in institutions on the other hand (Finger et al. 2010), where institutions must be understood as "*the rules of the game*" (North 1993: 12), and not as organizations. Specifically, innovations are acknowledged to happen as a result of interaction between institutional, technological and market actors, when institutions and technology are "misaligned". A graphical representation of the framework is available in Fig. 14.1.

When using this framework to look at the Seoul and Bogota cases, it seems that the development and implementation of the T-money and "Tu Llave" cards have taken completely opposite paths. In Seoul, the card was basically implemented to support a wide institutional bus reform (the Seoul Bus Reform) that aimed, among other things, at changing ownership of bus routes and implementing a citywide-

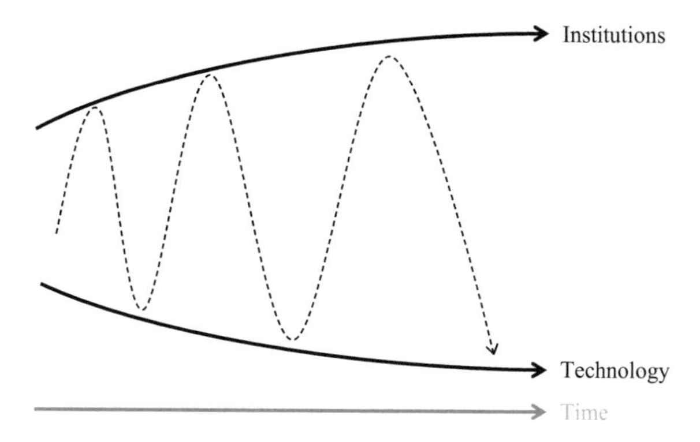

Fig. 14.1 The alignment framework (inspired from Finger et al. 2010)

integrated fare system. Thus, for the Seoul case, one can see the change having first happened on the institution axis of our framework, and then on the technology axis, because the Seoul bus reform created a misalignment that could only be tackled by the introduction of a technological innovation. The T-money card must thus be understood as a mean for public authorities to "catch-up" on institutional changes introduced, or at least as a way to support those. Although the card was actually developed by LG CNS prior to the bus reform, and even prior to the election of Mr. Lee Myung-Bak as the Mayor of Seoul (Lee and Lee 2013), its introduction was only made possible because of the introduction of institutional changes, known as the Seoul bus reform. In Bogota, things seem to have happened the other way around. The introduction of the smart card actually predated a wider institutional change in the transportation sector. Indeed, the "Tu Llave" card was introduced not as a mean to support a system-wide institutional reform (as it was the case in Seoul), but as a vector for the development of the SITP, which was being incrementally implemented as Bogota city hall was struggling to integrate pre-TransMilenio bus-operating companies into its new transportation network. Consequently, one can say that the introduction of the smart card in Bogota, that is to say, the technological change, might have happened a little bit too early, and not as a mean to catch-up on institutional change, thus perhaps making the institutional change (development of the SITP) more laborious to accomplish.

14.5 Conclusion

From this chapter, it seems that an IoT solution originally developed in the North can also produce positive outcomes when implemented in the South. Indeed, the implementation of an integrated smart ticketing system seems to have contributed to improving the experience of public transport users and the efficiency of the public transport system both for Seoul and Bogota. However, it also seems that, in order for southern cities to fully benefit from the technological innovation implemented, those must be implemented ex-post system-wide institutional reforms, and not preceding those. By implementing a technological solution "hastily", public authorities might indeed take the risk of delaying institutional changes at a system level, and consequently not fully harvest the potential of the technological solution. Building on Wright (2011), we thus think that for the case of Bogota, institutional changes (regulation of pre-TransMilenio operating companies) should have been introduced during the implementation of TransMilenio, and only after, the "Tu Llave" card should have been introduced, to consolidate the whole and back-up the institutional change introduced.

While we have seen in this paper that southern cities have some things to learn from northern cities regarding the implementation of IoT solutions, it also seems that southern cities might soon have some things to teach to northern cities. Indeed, over the last decade, the Information and Communication Technologies (ICTs) have enabled, as much in the South than in the North, the development, mainly by the private sector, of new mobility solutions, such as ride-booking, car-sharing, or car-pooling, that might enable to reduce private car use, and ultimately tackle congestion in urban areas (ITF 2017). Those new ICT-supported mobility solutions call for the development of new rules and regulations and some southern cities have actually been quite good at it. For example, innovative regulatory solutions developed towards ride-booking in Sao Paulo (Audouin and Neves 2017), might well become quickly inspirational for northern cities. Given those recent developments in urban mobility, it seems entirely relevant to look at transferability of solutions not only from a North to South perspective, but also from a South to North angle.

References

Allen, H. (2013). Bus reform in Seoul, Republic of Korea, in Global Report for human settlement 2013, UN-Habitat. http://www.unhabitat.org/grhs/2013. Accessed October 8, 2015.

Ardila, A. (2004). Transit Planning in Curitiba and Bogotá: Roles in Interaction, Risk, and Change. Ph.D. Dissertation. Cambridge: Massachusetts Institute of Technology.

Audouin, M., & Neves, C. (2017). *What regulations for ICT-based mobility services in urban transportation systems? The cases of ride-booking regulation in Sao Paulo and Rio de Janeiro. WIT Transactions on The Built Environment* (Vol. 176). Southampton and Boston: WIT Press.

Audouin, M., Razaghi, M., & Finger, M. (2015). *How Seoul used the 'T-Money' smart transportation card to re-plan the public transportation system of the city: Implications for governance of innovation in urban public transportation systems*. Paper Presented at the 8th TransIST Symposium in Istanbul, December 2015.

Bae, C. H., & Richardson, H. W. (2011). *Regional and urban policy and planning on the Korean Peninsula*. Cheltenham: Edward Elgar.

Blythe, P. T. (2004). Improving public transport ticketing through smart cards. *Proceedings of the Institution of Civil Engineers-Municipal Engineer, 157*(1), 47–54.

Bocarejo, J. P., Escobar, D., Hernandez, D. O., & Galarza, D. (2014). Accessibility analysis of the integrated transit system of Bogota. *International Journal of Sustainable Transportation, 24,* 142–154.

Business Korea. (2015). LG CNS Accelerates IT market invasion in Latin America. http://www.businesskorea.co.kr/english/news/ict/10187-colombian-success-lg-cns-accelerates-it-market-invasion-latin-america. Accessed November 17, 2017.

Cain, A., Darido, G., Baltes, M., Rodriguez, P., & Barrios, J. (2006). Applicability of Bogotá's TransMilenio BRT System to the United States, National Bus Rapid Transit Institute (NBRTI), FL-26-7104-01.

Cervero, R., & Kang, C. D. (2011). Bus rapid transit impacts on land uses and land values in Seoul Korea. *Transport Policy, 18*(1), 102–116.

Cervero, R., (2013). *Transport infrastructure and the environment: Sustainable mobility and urbanism.* Paper Presented at the 2nd Planocosmo International Conference, Bandung Institute of Technology, October 2013.

Dijkman, R. M., Sprenkels, B., Peeters, T., & Janssen, A. (2015). Business models for the Internet of things. *International Journal of Information Management, 35*(6), 672–678.

El Espectador. (2014). Integración de tarjetas del SITP en Bogotá terminará en 2015. http://www.elespectador.com/noticias/bogota/integracion-de-tarjetas-del-sitp-bogota-terminara-2015-articulo-518507. Accessed December 10, 2015.

Finger, M., Crettenand, N., & Lemstra, W. (2015). The alignment framework. *Competition and Regulation in Network Industries, 16*(2), 89–105.

Finger, M., Crettenand, N., Laperrouza, M, & Künneke, R. (2010). *Governing the dynamics of the network industries.* Discussion Paper Series on the Coherence Between Institutions and Technologies in Infrastructures, EPFL.

Heres, D., Jack, D., & Salon, D. (2013). *Do public transit investments promote urban economic development? Evidence from Bogotá*. Colombia: ITC—Institute of Transportation Studies, UC Davis.

Hillier, B. (2009, September). The city as a socio-technical system: A spatial reformulation in the light of the levels problem and the parallel problem. *Keynote Paper to the Conference on Spatial Information Theory*.

IFC. (2012). Recaudo Bogota, summary of investment information. http://ifcext.ifc.org/ifcext/spiwebsite1.nsf/ProjectDisplay/SII31907. Accessed on October 8, 2015.

ITF. (2017). *Shared mobility simulations for Helsinki*. OECD/ITF.

Kim, K. S., & Dickey, J. (2006). Role of urban governance in the process of bus system reform in Seoul. *Habitat International, 30*(4), 1035–1046.

Kim, S., & Shon, E. (2011). Effects of regulation changes in Seoul bus system: private bus operation under non-competitive fixed price contract. *Journal of advanced transportation, 45,* 107–116.

Kim, K. S., Cheon, S., & Lim, S. (2011). Performance assessment of bus transport reform in Seoul. *Transportation, 38*(5), 719–735.

Lara, A. Y., & Gutierrez, R. E. (2012). The Implementation of Integrated Transport System (SITP) of Bogotá and its challenges in the future. *Tecnogestión, 9*(1), 26–40.

Lee, J. H., & Lee, S. K. (2013). A Framework of the convergent service development process in the public sector: The smart transportation card service of Seoul City and the Call for collaboration case in Singapore. *Journal of the Korea society of IT services, 2*(2), 387–410.

McHale, M., Pickett, S., Barbosa, O., Bunn, D., Cadenasso, M., Childers, D., et al. (2015). The new global urban realm: Complex, connected, diffuse, and diverse social-ecological systems. *Sustainability, 7*(5), 5211–5240.

Munoz-Raskin, R. (2010). Walking accessibility to bus rapid transit: Does it affect property values? The case of Bogota. *Transport Policy, 17*(2), 72–84.

North, D. C. (1993). Institutions and credible commitment. *Journal of Institutional and Theoretical Economics, 149*(1), 11–23.

Park J. Y., & Kim D. (2013). Korea's integrated fare and smart cart ticket system: Innovative PPP (Public-Private Partnership) approach. KOTI Knowledge Sharing Report, Issue 05.

Pucher, J., Park, H., Han Kim, M., & Song, J. (2005). Public transport reforms in Seoul: Innovations motivated by funding crisis. *Journal of Public Transportation, 8*(5), 41–62.

Ramos, C. (2015), Public transport in Colombia: transition period towards the implementation of the integrated public transport system of Bogota (SITP). *Master Thesis*, Cardiff University.
SDP. (2009). Estudio prospectivo del Sistema Integrado de Transporte de Bogotá y la Región. http://www.sdp.gov.co/portal/page/portal/PortalSDP/InformacionTomaDecisiones/Documental/consultaDocumentos/Estudio_prospectivo_movilidad.pdf. Accessed November 17, 2017.
SMG. (2014). Seoul public transportation. https://citynet-ap.org/wp-content/uploads/2014/06/Seoul-Public-Transportation-English.pdf. Accessed November 17, 2017.
UN-Habitat. (2013). State of the world's cities 2012/2013, prosperity of cities. http://mirror.unhabitat.org/pmss/listItemDetails.aspx?publicationID=3387. Accessed October 8, 2015.
World Bank. (2014). Bus concession contracts and tariff policy: Lessons from the Bogotá and Colombia experience. http://docs.trb.org/prp/15-2951.pdf. Accessed December 10, 2015.
World Bank. (2010). From Chaos to order implementing high-capacity urban transport systems in Colombia. http://documents.worldbank.org/curated/en/100571468244175261/pdf/916790BRI0Box30nsit402301000Public0.pdf. Accessed November 17, 2017.
Wright, L. (2011). Bus rapid transit: A review of recent advances. In H. Dimitriou & R. Gakenheimer (Eds.), *Urban transport in the developing world: a handbook of policy and practice* (pp. 421–455). Cheltenham: Edgar Elward.
Zanella, A., Bui, N., Castellani, A., Vangelista, L., & Zorzi, M. (2014). Internet of things for smart cities. *IEEE Internet of Things Journal, 1*(1), 22–32.

Part VI
Disaster Risk Reduction

Chapter 15
Putting 200 Million People "on the Map": Evolving Methods and Tools

Emily Eros

15.1 Introduction

Accurate maps play a critical role in understanding human communities, particularly for populations at risk. Maps help individuals to understand their surroundings, locate features of interest, plan routes and journeys, and conduct broader scale processes such as urban planning and disaster response. In crisis situations specifically, basemaps and geospatial data act as key tools for first responders and NGOs to visualize damage assessments; assess infrastructure, hazards, and demographics; plan local and larger scale emergency response activities; and share information among relevant actors. A lack of maps and geographic data can prevent lifesaving resources from meeting those in need. For this reason, mapping is a core component of the Red Cross' disaster response efforts.

Much of the Global North has been mapped with incredible detail. With an Internet connection or a smartphone, individuals can instantly access detailed address information, building footprints, transit information, and locations for health facilities, public services, and businesses. This is still not the reality for billions of people in vulnerable areas, however. Popular services such as Google Maps lack information for much of the Global South; many communities exist as simply a place name and perhaps a collection of roads. Other communities are absent entirely. Even when a community is mapped at a detailed level, information published without an open license (e.g., many web-maps and PDF maps) offers limited value to humanitarian actors. A user often cannot download the information, combine it with other datasets, and create and share products from it. Thus, the user is essentially viewing a picture of data rather than accessing the data directly. The same is true for mapping products created by government agencies and other traditional mapping authorities, who have

E. Eros (✉)
American Red Cross, Washington, DC, USA
e-mail: emily.eros@gmail.com

S. Hostettler et al. (eds.), *Technologies for Development*,
https://doi.org/10.1007/978-3-319-91068-0_15

traditionally operated on a retail model of distribution and often do not release their data for free and open use (Goodchild and Glennon 2010).

OpenStreetMap (OSM) has greatly improved this situation for humanitarians. Often described as "the Wikipedia of maps" (Heinzelman and Waters 2010), OSM is a free online map and geospatial database whose data can be viewed, edited, downloaded, and used by anyone with an Internet connection. While there are several other geographically focused examples of what academics refer to as Web 2.0 (Vossen and Hagemann 2007), crowdsourcing (Howe 2008), or volunteered geographic information (Goodchild 2007), OSM is the most prominent platform, with a community of over two million registered users (OSM Contributors 2016). Data quality and quantity vary by region, but previous studies of locations such as the UK (Haklay 2010), Ireland (Ciepłuch et al. 2010), France (Girres and Touya 2010), and Iran (Forghani and Delavar 2014) confirmed that the data are generally of high quality.

OSM plays an important role for many humanitarian organizations as their primary source for base map data and primary platform for new data creation. At the time of the 2010 Haiti earthquake, for example, only rudimentary maps were widely available. Within hours of the earthquake, over 600 OSM users around the world began rapidly tracing satellite imagery to create a detailed map of the country, which disaster response and relief agencies used as an important tool in relief and recovery efforts—both in the immediate aftermath and in the years to come (Soden and Palen 2014; Zook et al. 2010). A non-profit organization called the Humanitarian OpenStreetMap Team (HOT) was formed as a result. HOT coordinates the creation and distribution of mapping resources to assist humanitarians responding to crises and disasters around the world (Neis and Zielstra 2014). The group played a key role in organizing similar mapping efforts after the 2013 Typhoon Haiyan in the Philippines (Palen et al. 2015), the 2014 West African Ebola outbreak (Cassano 2014), the 2015 Nepal earthquake (Sneed 2015), and countless other small-scale disasters and for communities at risk.

In addition to post-disaster mapping, OSM has also acted as a platform for mapping disaster-prone communities in vulnerable areas. Map Kibera led to the grassroots mapping of Kenya's biggest slum community (Hagen 2011). HOT launched a large-scale community-based effort to map Jakarta and other parts of earthquake-prone Indonesia (Soden and Palen 2014). A World Bank Open Cities project prompted detailed mapping of the central Kathmandu Valley, which led to the creation of a local non-profit, Kathmandu Living Labs (Ibid.). These efforts have been highly successful in building communities of skilled, engaged OSM contributors who generated valuable data, which played a key role to responders following floods in Indonesia and the 2015 Nepal earthquake.

Despite these successes, the communities home to billions of people remain "off the map". Because of the digital divide and differences in levels of development, geographic information "is usually least available where it is most needed" (Sui et al. 2013: 5). For the humanitarian community, further efforts are needed to preemptively generate baseline data before major crises occur. Recognizing this need, the American Red Cross, in partnership with the Humanitarian OpenStreetMap Team (HOT), the British Red Cross, and Médecins Sans Frontières UK, launched an innovative project called Missing Maps, which aims to put 200 million vulnerable people on the map by 2021—adding the communities in which they live to OSM, where the data are openly available for use and revision. This represents a targeted effort to conduct widespread pre-disaster mapping without limiting the focus to a particular geography.

This paper (1) describes the Missing Maps project's remote and field mapping methods, (2) outlines the technical tools developed for the project, (3) highlights how humanitarian and development organizations can adapt these tools and methods to other initiatives, and (4) explores current challenges and research questions surrounding mapping initiatives focused on the Global South. In doing so, this paper provides an overview of current trends in crowdsourced mapping and emerging data collection methods, with the aim to share tools and experiences with others in the humanitarian community.

15.2 Remote and Field Mapping Methods

As of November 2017, three years after the project's launch, Missing Maps volunteers collectively mapped an estimated 60 million people. The majority of this progress involved remote mapping efforts. The American Red Cross and its partners host and support mapping parties ("mapathons") in cities around the world, often working with university groups or even private corporations who are interested in bringing people together and training them how to trace buildings and roads from satellite imagery in order to map a vulnerable area.

This engagement strategy is modeled after mapping parties and "crisis camps" held by HOT, Kathmandu Living Labs (KLL), the World Bank, and other organizations (see Soden and Palen 2014). Missing Maps is unique, however, in its unprecedented number of events and its geographic reach. Figure 15.1 shows the locations and density of Missing Maps mapathons as well as the location of every building mapped by Missing Maps volunteers from the project's launch in November 2014 until the time this article was written, in November 2017. Partners have hosted a total of 623 public mapathons across all six continents, resulting in nearly 40,000 attendees mapping over 15 million buildings and 13 million kilometers of roads in OSM. Outside of mapping parties, volunteers also remotely contribute to mapping tasks. Missing Maps continues to build momentum, with 17 organizations now part of the partnership.

Remote tracing efforts very quickly result in basemap data for the area of interest. Over a matter of days, a community can go from a blank spot on the map to a detailed collection of buildings and roads. This gives humanitarian agencies necessary data to plan and organize their community development and disaster preparedness/response activities.

Remote mapping rapidly produces data but faces certain limits. Building and road outlines provide a baseline of physical infrastructure information for humanitarians, but additional information is necessary to make the map more complete and useful. It is impossible to tell from satellite imagery where hospitals, schools, water and

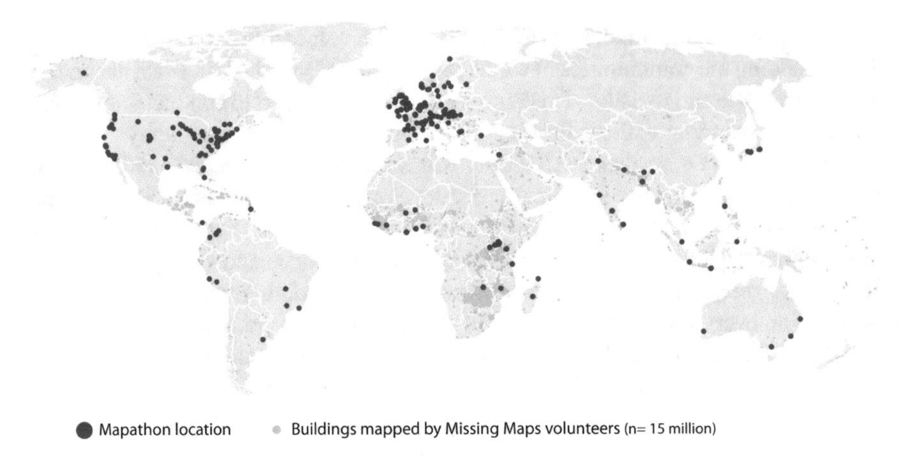

Fig. 15.1 Missing maps mapathons and target mapping areas, as of November 2017

sanitation facilities, community landmarks, and physical hazards are located. Moreover, local community engagement is important in order to promote sustainability and a sense of ownership over the map data. Therefore, the American Red Cross began to increasingly focus on the local engagement and field-based activities.

The American Red Cross had led Missing Maps field training in countries such as Bangladesh, Belize, Colombia, Ecuador, Guinea, Jamaica, Liberia, Rwanda, Sierra Leone, South Africa, Tanzania, and Zimbabwe. For each field visit, remote volunteers map the area of interest, then a team of staff visit the local area to work with local Red Cross volunteers in order to explain about the mapping project, train the volunteers in necessary processes and technology, and organize the fieldwork. The team consults local officials and Red Cross staff and volunteers to solicit their input on data collection priorities, which often include water points, waste disposal sites, and building or road conditions. Staff mentor the volunteers to help them edit and upload the resulting data. After the project, they leave equipment behind with the local Red Cross in order to enable future mapping efforts. Because areas of interest are selected based on current project locations and needs, the data add immediate value to resilience projects, community assessments, and other ongoing activities.

Short field projects provide a valuable opportunity to engage local communities and add contextual data to the map. However, engagement based on bursts of international travel poses challenges in terms of sustainability. It can be very difficult for organizations to foster long-term engagement and support local mappers without an embedded presence in the community. Past experience suggests that, even when local volunteers received extensive training and staff left equipment behind for the volunteers' use, limited mapping activities continued independently after the site visits concluded. This prompted the team to consider additional strategies to increase community engagement and sustainability.

As one response to these challenges, the American Red Cross launched a more embedded approach in West Africa, where it launched a mapping hub in Guéckédou,

Guinea—the epicenter of the 2014 Ebola outbreak. The hub operated in 2015–2016 as a physical office location with equipment, four technical staff, administrative support, and training facilities. It acted as a base for specific mapping projects and to host community meetings, conduct outreach events, and offer training modules designed to boost basic computer skills and build up a large community of volunteers with skills in various mapping and data-gathering technologies. As volunteers become more experienced, they played a larger role in leading these activities.

The hub initially supported the rapid mapping of all communities (totaling 7000) within 15 km of the international borders shared between Liberia, Guinea, and Sierra Leone—focusing on water, sanitation, health, and public amenities in each community. Volunteers on motorbikes carried out the fieldwork, equipped with a smartphone app (OpenMapKit) and other technical tools. Resulting data were published on OSM and the Humanitarian Data Exchange, and were shared with decision makers, humanitarian workers, and community stakeholders to ensure they are well aware of water, sanitation, health, and community resources before and during the next regional crisis—helping them to make data-driven programming decisions and potentially preventing or slowing the spread of infectious disease during future disease outbreaks.

Although the West Africa project has ended and the hub has closed, equipment was donated to nascent OSM groups who continue to conduct mapping and fieldwork; in Guinea, for example, the local mapping community conducted 10 mapathons to support local mapping as well as disaster response activations for Hurricane Irma, malaria prevention in Zambia, and a recent earthquake in Mexico. Over 300 community members have joined OSM Guinea and a further 400 have joined OSM Liberia, suggesting that the investment into mapping has spurred a lasting interest in community-based, participatory mapping.

Moving forward, the American Red Cross and its partners continue to pursue opportunities for sustained engagement, and will continue to evaluate their methods in order to respond to challenges and promote best practices.

15.3 Technical Tools

For remote mapping, Missing Maps trains volunteers in OSM using an in-browser editing interface (iD Editor), which most mapathon participants have easily learned within a few minutes of instruction. The team also produces training videos and resources for mapping that provides a reference and examples for volunteers.[1] As volunteers gain experience, we encourage the use of the Java OSM editor (JOSM), which is a faster and more robust editing interface but can also be more difficult for beginners to learn.

Beyond these existing tools, the American Red Cross introduced new efforts to gamify remote mapping and keep trained volunteers engaged between disasters.

[1] Examples are embedded into mapping tasks, available at http://www.tasks.hotosm.org.

The team developed an online tracking tool and API that compiles stats and metrics for each Missing Maps volunteer, which feeds into user profile webpages with badges for different achievements and "levels" the mapper has reached.[2] In addition, the tracking tool may help the Missing Maps team to identify dedicated and experienced remote volunteers who may be interested in hosting their own mapping events in their communities in the future.

A combination of high- and low-tech tools facilitate field mapping. The team's primary tool is OpenMapKit,[3] an Android app the American Red Cross developed that enables a user to pre-download the OSM map of an area, shows the user's location, and enables him/her to simply tap on a building or feature in order to add information about it. When an object is tapped, a brief set of questions will appear, asking the user what sort of object this is (e.g., school, residential building, health facility) and for any other relevant information. These fields are determined based on community priorities and created at the outset of the mapping project, using OpenDataKit forms. Volunteers upload their mapping results from the mobile devices, and the data are added to OSM, where staff review the additions for quality control. Teams can easily map thousands of buildings in a week using this method.

When the use of mobile phones presents security concerns, or other factors necessitate an alternative method, the American Red Cross uses paper-based mapping tools (such as Field Papers, a tool which enables users to print map pages for an area, annotate the pages by hand, and then scan/edit/upload the annotated information to OpenStreetMap). These tools are freely available and easy to use.

In addition, the American Red Cross developed new tools to facilitate OSM editing in areas with unreliable connectivity. Humanitarians, in general, have become increasingly reliant on cloud-based tools for use in increasingly remote areas. For mapping projects specifically, users previously needed an Internet connection to use OpenMapKit in order to download map areas at the beginning of a project, and then to upload data into OSM at the end of each day. Field Papers also requires an Internet connection to prepare map atlases. Reliable Internet access on a daily basis is not realistic for many areas in which the Red Cross and other humanitarians work.

With this in mind, the American Red Cross created a new tool: Portable OSM (POSM). POSM combines an offline OSM API, Field Papers, OpenMapKit, a captive portal, and many new enhancements to the OSM workflow. POSM runs on very inexpensive hardware (costing roughly $300 USD) and free, openly available software. With POSM, users are able to download all necessary files for an area of interest, easily configure the portable server, work offline in the field (using OpenMapKit, OpenDataKit, iD Editor, JOSM, and OpenDroneMap), and then return to a connected environment and sync all changes back to OpenStreetMap. With this new technology, users can conduct mapping or mobile data collection fieldwork for upwards of a month at a time without connecting to the Internet. User-friendly documentation is available at the POSM website[4] and full source code is available online.

---

[2] Available at http://missingmaps.org/users.

[3] Available at http://www.openmapkit.org.

[4] Available at http://www.posm.io.

15.4 Applications and Potential Development Impacts

Missing Maps' intent is to enable local communities and humanitarian organizations to collect useful geospatial information. Accordingly, the field mapping methods described above use low-cost devices (basic smartphones), rely on applications that are easy to learn, and use free and open-source software. All of the technical tools mentioned in the previous section are freely available for public use. This includes tools developed by others (iD editor,[5] JOSM,[6] OpenDataKit,[7] Field Papers[8]) and those developed by the American Red Cross (OpenMapKit, POSM). Various training materials exist for the former, and instructional materials were to be created for the latter. Source code can also be found on the American Red Cross' GitHub[9] page. In addition, Missing Maps partner organizations have published materials to support individuals interested in hosting a mapathon,[10] and the team regularly provides support to individuals who contact them with questions.

Missing Maps' field mapping methods and tools can be leveraged by the humanitarian/development community for a variety of situations. For example, an organization could use OpenMapKit specifically to create a basemap of the area in which it is working, to monitor the state of water points and sanitation facilities, to assess damaged buildings and roads after a natural disaster, to create a database of school locations and addresses, etc. POSM can be used by small mapping teams to map large areas with variable connectivity; it can bridge connected and disconnected environments and provide an offline backup for intermittent Internet connectivity.

15.5 Challenges and Research Directions

By engaging remote mappers and local volunteers to map 60 million people in three years, the Missing Maps project and its partner organizations have made significant contributions to tools and methods for mapping vulnerable communities. Based on its experiences, the American Red Cross has highlighted certain challenges and areas for future growth.

For remote mapping, an important challenge is developing new ways to keep volunteers engaged between disasters. High numbers of volunteers express interest in remote tracing after a major disaster. However, remotely training a large pool of extremely diverse volunteers requires time, proper tools, and coordination. A disaster setting is not the ideal time to start the process. Instead, it is imperative to have a large body of well-coordinated volunteers that are ready to assist when a disaster occurs.

---

[5] Available at http://www.openstreetmap.org/edit?editor=id.

[6] For more information: http://wiki.openstreetmap.org/wiki/JOSM.

[7] For more information: https://opendatakit.org.

[8] For more information: http://fieldpapers.org.

[9] https://github.com/AmericanRedCross.

[10] For more information: http://www.missingmaps.org/mapathons.

The American Red Cross is actively developing digital tools and ways to gamify remote mapping, but the humanitarian community could benefit from additional insight and tools for engaging remote volunteers. Additionally, ensuring data quality becomes especially important when adding large numbers of new mappers into the OSM community. The American Red Cross has emphasized the importance of data validation and held events for more experienced mappers in order to check and correct OSM data. More work remains in this area.

Machine learning algorithms offer another area for future research and exploration. For larger mapping needs where crowdsourcing is less feasible or timely, organizations recently began to experiment with machine learning methods and outputs. For example, Facebook, Columbia University, and the World Bank pioneered a method to identify buildings from high-resolution,[11] commercially available satellite imagery (Tiecke 2016). The partners trained the model to adapt it to local conditions in 18 different countries, ground truthed the results against household survey data and existing crowdsourced OSM building data, and applied most recent government census population data to outputs. The result is openly accessible, nationwide datasets that contain estimates of human population down to the 30-m scale. Data are available for 18 countries at the time of publication (Facebook Connectivity Lab and CIESIN 2016) and agencies like the Red Cross have begun testing their applications for humanitarian purposes, and in combination with crowdsourced and locally derived data. As these techniques become even more refined, more work is necessary to investigate how computer algorithms, crowdsourcing, and local knowledge can work together to create valuable data for humanitarian purposes while also respecting privacy and sensitivities.

For field mapping, sustainability and long-term engagement are areas for additional growth—especially for organizations working internationally. There have been highly successful initiatives undertaken by HOT in Indonesia and by the World Bank/KLL in Kathmandu, and mapping engagement has continued following the end of the American Red Cross' mapping project in West Africa. Again, this area could benefit from additional insight into ways to effectively engage with local communities, promote local interest and ownership over data collection, and support a community of mappers, who are trained and motivated to continue mapping activities independently. Rural areas may present a particular challenge; previous tech hubs and mapping initiatives have been based out of urban centers like Nairobi, Kathmandu, and Jakarta, which benefit from more developed infrastructure, the presence of university students, and a more technically skilled volunteer base.

Finally, as the mapping community grows, it is important to bridge the gap between those who are experienced in working with data and technology, and the broader community of practitioners and decision makers, who may be less comfortable with digital tools and resources. Base data and analytics bring higher value when more people understand how to interpret and apply them to situations. For this reason, the Missing Maps partnership has emphasized efforts to increase data literacy and what we call,

[11]0.5 m resolution.

"data readiness"—the ability of an organization to utilize data in response operations for situation awareness, planning, implementation, monitoring, and reporting.

As the Missing Maps project moves forward, the American Red Cross and its partner organizations will continue to evaluate their methods and tools for mapping, evolving new tools and engagement strategies to address the challenges experienced along the way. By mapping vulnerable communities, humanitarians are better able to share better analysis of hazards, mitigations, and response capacity in communities, enabling stronger disaster response efforts, and program planning in the future.

References

Cassano, J. (2014, October 22). *Inside the crowdsourced map project that is helping contain the Ebola epidemic*. Fast Co. Labs. Retrieved from: http://www.fastcolabs.com/3037350/elasticity/inside-the-crowdsourced-map-project-that-is-helping-contain-the-ebola-epidemic.

Ciepłuch, B., Jacob, R., Mooney, P., & Winstanley, A. (2010). Comparison of the accuracy of OpenStreetMap for Ireland with Google Maps and Bing Maps. In *Proceedings of the Ninth International Symposium on Spatial Accuracy Assessment in Natural Resources and Environmental Sciences, July 20–23rd, 2010*. Leicester: University of Leicester.

Facebook Connectivity Lab and Center for International Earth Science Information Network (CIESIN), Columbia University. (2016). *High resolution settlement layer (HRSL)*. Retrieved November 6, 2017 from: https://ciesin.columbia.edu/data/hrsl/.

Forghani, M., & Delavar, M. R. (2014). A quality study of the OpenStreetMap dataset for Tehran. *ISPRS International Journal of Geo-Information, 3*(2), 750–763.

Girres, J. F., & Touya, G. (2010). Quality assessment of the French OpenStreetMap dataset. *Transactions in GIS, 14*(4), 435–459.

Goodchild, M. F. (2007). Citizens as sensors: the world of volunteered geography. *GeoJournal, 69*(4), 211–221.

Goodchild, M. F., & Glennon, J. A. (2010). Crowdsourcing geographic information for disaster response: a research frontier. *International Journal of Digital Earth, 3*(3), 231–241.

Hagen, E. (2011). Mapping Change: Community Information Empowerment in Kibera (Innovations Case Narrative: Map Kibera). *Innovations, 6*(1), 69–94.

Haklay, M. (2010). How good is volunteered geographical information? A comparative study of OpenStreetMap and Ordnance Survey datasets. *Environment and Planning B: Planning and Design, 37*(4), 682–703.

Heinzelman, J., & Waters, C. (2010). *Crowdsourcing crisis information in disaster-affected Haiti*. Washington, DC: US Institute of Peace.

Howe, J. (2008). *Crowdsourcing: Why the power of the crowd is driving the future of business*. New York: McGraw-Hill.

Neis, P., & Zielstra, D. (2014). Recent developments and future trends in volunteered geographic information research: The case of OpenStreetMap. *Future Internet, 6*(1), 76–106.

OpenStreetMap Contributors. (2016). *Wiki: Stats*. Retrieved January 13, 2016 from: http://wiki.openstreetmap.org/wiki/Stats.

Palen, L., Soden, R., Anderson, T. J., & Barrenechea, M. (2015, April). Success & scale in a data-producing organization: the socio-technical evolution of OpenStreetMap in response to Humanitarian events. In *Proceedings of the 33rd Annual ACM Conference on Human Factors in Computing Systems* (pp. 4113–4122). New York: ACM.

Sneed, A. (2015, May 8). The open source maps that make rescues in Nepal possible. *Wired*. Retrieved from: http://www.wired.com/2015/05/the-open-source-maps-that-made-rescues-in-nepal-possible/.

Soden, R., & Palen, L. (2014). From crowdsourced mapping to community mapping: The post-earthquake work of openstreetmap Haiti. In *COOP 2014: Proceedings of the 11th International Conference on the Design of Cooperative Systems*, 27–30 May 2014 (pp. 311-326). Nice, France: Springer International Publishing.

Sui, D., Goodchild, M., & Elwood, S. (2013). Volunteered geographic information, the exaflood, and the growing digital divide. In *Crowdsourcing Geographic Knowledge* (pp. 1–12). Netherlands: Springer.

Tiecke, T. (2016). *Open population datasets and open challenges*. Facebook Connectivity Lab. Retrieved November 6, 2017 from: https://code.facebook.com/posts/596471193873876.

Vossen, G., & Hagemann, S. (2007). *Unleashing Web 2.0: From Concepts to Creativity.* Boston: Elsevier/Morgan Kaufmann.

Zook, M., Graham, M., Shelton, T., & Gorman, S. (2010). *Volunteered geographic information and crowdsourcing disaster relief: a case study of the Haitian earthquake*. Available at SSRN: http://ssrn.com/abstract=2216649.

Chapter 16
Highlights and Lessons from the Implementation of an Early Warning System for Glacier Lake Outburst Floods in Carhuaz, Peru

Javier Fluixá-Sanmartín, Javier García Hernández, Christian Huggel, Holger Frey, Alejo Cochachin Rapre, César Alfredo Gonzales Alfaro, Luis Meza Román and Paul Andree Masías Chacón

16.1 Introduction

The Peruvian high mountains are often affected by mountain disasters such as ice and rock avalanches, Glacial Lake Outburst Floods (GLOFs), floods, or debris flows. Among them, outbursts of glacier lakes are considered as the most far-reaching glacial hazard (Kääb et al. 2005). In particular in relatively densely populated mountain ranges such as the European Alps, the Himalayas, or the tropical Andes, where

J. Fluixá-Sanmartín (✉) · J. García Hernández
Centre de Recherche sur l'Environnement Alpin (CREALP), Sion, Switzerland
e-mail: javier.fluixa@crealp.vs.ch

J. García Hernández
e-mail: javier.garcia@crealp.vs.ch

C. Huggel · H. Frey
Department of Geography, University of Zurich (UZH), Zurich, Switzerland
e-mail: christian.huggel@geo.uzh.ch

H. Frey
e-mail: holger.frey@geo.uzh.ch

A. Cochachin Rapre
Unidad de Glaciología y Recursos Hídricos (UGRH), Autoridad Nacional de Agua, Huaraz, Peru
e-mail: jcochachin@ana.gob.pe

C. A. Gonzales Alfaro
CARE Perú, Huaraz, Peru
e-mail: cgonzales252006@gmail.com

L. M. Román
Municipalidad de Carhuaz, Carhuaz, Peru
e-mail: arq.luchomeza@gmail.com

P. A. Masías Chacón
Corporación RD S.R.L, Cusco, Peru
e-mail: paulmasias@gmail.com

S. Hostettler et al. (eds.), *Technologies for Development*,
https://doi.org/10.1007/978-3-319-91068-0_16

infrastructure and settlements are located within the range of potential GLOFs, the risk emanating from glacier lakes has to be addressed.

As a major element of an integrated risk management strategy, Early Warning Systems (EWSs) represent a viable and promising nonstructural tool for mitigating climate change-related risks (Bulmer and Farquhar 2010; Huggel et al. 2010). It prevents loss of life and reduces the economic and material impact of disasters. To be effective, EWS needs to actively involve the communities at risk, facilitate public education and awareness of risks, effectively disseminate messages and warnings, and ensure there is constant state of preparedness (ISDR 2006).

Four main interlinked elements of an EWS can be identified (ISDR 2006): (i) Risk Knowledge through the collection and analysis of data concerning hazards and vulnerabilities, evacuation routes, etc.; (ii) Monitoring and Warning Service, at the core of the system, for predicting and forecasting hazards, and for continuously monitoring hazard parameters, which is essential to the generation of accurate warnings in a timely fashion; (iii) Dissemination and Communication, for warnings must reach those at risk using predefined national, regional, and community appropriate communication systems; and (iv) Response Capability where education and preparedness programs play a key role.

EWSs in high-mountain contexts are highly complex systems (Frey et al. 2014; Schneider et al. 2014). On the one hand, they have to include monitoring sensors and a communication network for data and voice. On the other hand, they have to establish clear procedures, define institutional responsibilities and response measures, and most importantly involve local stakeholders to ensure that adequate actions are taken according to different warning levels.

In this paper, the main features of the EWS implemented in the Carhuaz region and the relevant lessons learned from the project are presented.

16.2 Study Region

The Cordillera Blanca, in the tropical Andes of Peru (Fig. 16.1), supplies water to many towns and cities located in the valley of the Callejon de Huaylas. It is however also a source of hazards due to the occurrence of ice and rock avalanches, and ensuing GLOFs which have been historically threatening the population of this area (Carey et al. 2012). More recently, in April 2010, an overflow of the glacial lake "Laguna 513" (Huaraz, Ancash) was caused by the impact of an avalanche of rock and ice, which in turn triggered a flood wave that transformed into a debris flow that impacted downstream areas reaching the city of Carhuaz. Fortunately, nobody was seriously injured, but damage to property was considerable.

This mountain range has a glacier coverage of more than 500 km^2 (Racoviteanu et al. 2008), which accounts for about 25% of the world's tropical glaciers. The glacial lake "Laguna 513" (4428 m a.s.l., 9° 12′ 45″S, 77° 33′ 00″W) formation started in the early 1970s by filling a basin that was uncovered by the shrinking Glacier 513, and several smaller lake outbursts took place in the 1980s. In the early 1990s, the

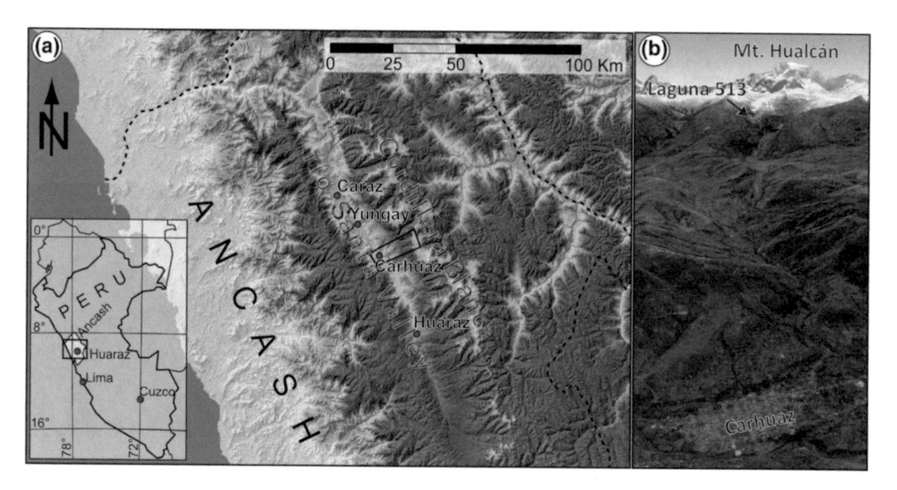

Fig. 16.1 **a** Location of the Cordillera Blanca, Peru (Schneider et al. 2014). Black rectangle indicates the location of b. **b** Oblique view of Mt. Hualcán, Laguna 513, and the city of Carhuaz in the foreground (GoogleEarth)

lake level was lowered artificially, and a permanent tunnel system was installed in the bedrock dam of the lake, a structural measure that reduces not only the probability of failure but also the social and economic consequences.

Laguna 513 is one of the 14 lakes declared as posing a high hazard in the Cordillera Blanca (ANA 2014) endangering the growing population of the Callejón de Huaylas. Due to its vicinity to the settlements, combined with the frequent avalanching and the seismic activity, an integrated hazard assessment, including a detailed study of the current hazard situation and possible future scenarios as well as measures to improve local knowledge, capacities, and specific adaptation measures, was urgently required, as emphasized in different studies (Mark and Seltzer 2005; Racoviteanu et al. 2008; Vuille et al. 2008).

From 2011 to 2015, a multidisciplinary project for sustainable reduction of climate change-related risk in high mountains was accomplished: "Glaciares 513—Adapting to climate change and reducing disaster risks due to receding Andes glaciers". Within the context of this project, an operational EWS against GLOFs was designed and implemented at Laguna 513, and the necessary emergency response protocols were developed (Muñoz et al. 2015). CARE Peru's experience implementing community EWS, and the University of Zurich with a large experience in glacier hazard and risk research, facilitated the design of this innovative and unique system in Latin America.

Fig. 16.2 Combined hazard levels from the three scenarios; *lower panel*: generalized form, corresponding to the final hazard map for GLOFs for the entire catchment (Schneider et al. 2014)

16.3 Description of the EWS Implemented in Carhuaz

16.3.1 Risk Knowledge (GLOF Modeling)

In order to assess the stability conditions of the Mount Hualcán and the response of Laguna 513 facing potential avalanches, the modeling approach used by Schneider et al. (2014) to reconstruct the 2010 outburst was used. This model chain represents the process as an iterative cascade of interacting physically based models. The model was used to simulate both the 2010 outburst and an ensemble of potential scenarios of different magnitudes, which finally resulted in a hazard map for GLOF hazards for the entire catchment, including the urban area of Carhuaz.

For the modeling, the process chain of the 2010 event was divided into three main parts: the rock-ice avalanche, the displacement wave in the lake, and the GLOF, which was further subdivided into the different flow types. For this purpose, two models were used: the hydrodynamic IBER model (IBER, 2010) for the spillover hydrograph definition and the displacement wave, and the RAMMS (RApid Mass MovementS) model (Christen et al. 2010) for the avalanche and GLOF modeling. Inundated areas modeled by RAMMS correspond well with field evidences and post-event imagery.

According to the guidelines from Raetzo et al. (2002), flow velocities and heights can be translated into different intensity levels. Three scenarios were defined: the dimensions of the 2010 event were taken for the small scenario (450,000 m^3); the medium and large scenarios involved avalanche volumes of 1 and 3 million m^3, respectively (Schneider et al. 2014). Combining those levels of intensity with the flow heights and velocities modeled for the three scenarios, the final hazard map was obtained (Fig. 16.2). This map was then used to define the evacuation routes that were assessed and approved by the civil defense platform.

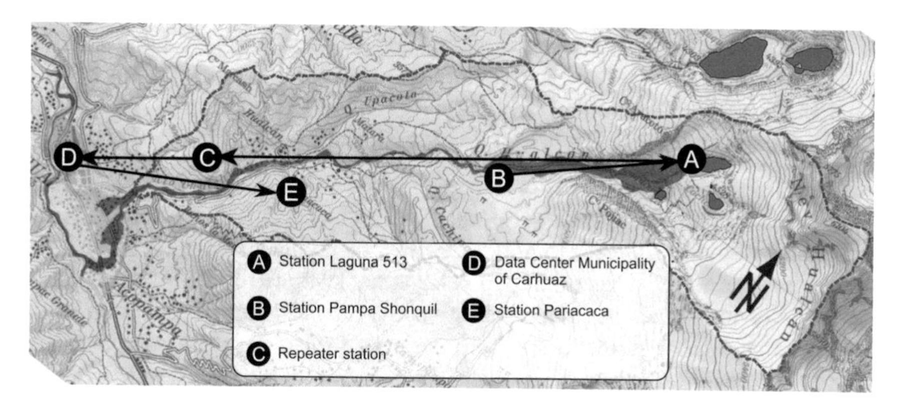

Fig. 16.3 The five stations of the EWS, arrows indicating the direction of the signal transfer. Background: topographic map from the Austrian Alpenverein and the GLOF hazard map (from Frey et al. 2014)

16.3.2 Monitoring and Warning Service (Implementation of Stations and Sensors)

The monitoring and warning system installed in Carhuaz comprises the following stations and sensors (Fig. 16.3):

A. **Glacier and lake monitoring station at Laguna 513 (4491 m a.s.l.)**: four geophones (devices that record ground movements due to avalanches on Laguna 513 and convert them into voltage) mounted in the bedrock and located close to the station, measuring continuously and sending data every 5 s; additionally, two cameras taking photos every 5 s during daylight hours: one oriented to the face of Mt. Hualcán, the other to the dam of Laguna 513. The station receives the signals from the geophones, the cameras and from the station at Pampa Shonquil, stores them in a data logger and broadcast them to the repeater station.
B. **Hydrometeorological monitoring station at Pampa Shonquil (3600 m a.s.l.)**: it includes a pressure sensor installed in the Chucchún River for monitoring river water level, as well as a meteorological station with sensors for measuring air temperature and humidity, precipitation, wind speed, and solar radiation. It stores the information in a data logger and sends the signal to the station at Laguna 513, since there is no direct view contact with the repeater station (C).
C. **Repeater station (3189 m a.s.l.)**: for transferring the signal from the Laguna 513 station to the data center (D). It receives the signals from the station at Laguna 513 and broadcast them to the data center in Carhuaz.
D. **Data center, located in Carhuaz in the building of the municipality (2640 m a.s.l.)**: it receives all information collected by the monitoring stations in real time. With this information, the mayor can take the decision to evacuate the population. It consists of a receiving antenna, a screen with real-time data access, and a server for data storage.

E. **Warning station located in the community of Pariacaca (3053 m a.s.l.)**: this station consists of monitoring equipment that allows the community to view real-time data of the Laguna 513. This station has two sirens that are activated from the data center (D) when a GLOF is confirmed.

All stations were equipped with solar panels and batteries for energy generation and storage, and had a mast where most of the instruments are fixed, a concreted and lockable box for the electronic equipment, and a protection fence. Energy availability is a critical and limiting factor, in particular at the station at Laguna 513, because the peaks of the Cordillera Blanca experience a much higher frequency in cloud coverage that regions further away from the main peaks. For preventing data loss and interrupted access in case of blackouts, emergency power batteries were available in the building of the municipality.

While the geophones are the principle instruments to register a potential GLOF trigger, the cameras are used as a backup as well as means for viewing the current situation, and, in particular during the test phase of the system, for relating geophone measurements to the magnitude of (avalanche) events. The pressure sensor in the riverbed at the Pampa Shonquil station adds redundancy to the system on the one hand and, if calibrations measurements are taken, can be used for continuously monitoring the runoff. Next to the station at Pampa Shonquil, there is a permanently manned hut of the wardens of the freshwater intake of Carhuaz. The warden is instructed to notify the authorities in case of an event, which is an additional and essential component of redundancy to the system.

16.3.3 Dissemination and Communication (Data Management)

All recorded data is stored first in the data logger at the respective station, then after data transmission (5 s intervals), on a server located in the data center and backed up on a cloud service. All the data can be assessed through a website to allow for real-time remote monitoring. In the data center itself—a dedicated office in the Municipality of Carhuaz—a screen displays the data from this web page continuously (24/7). The web page is structured in nine sub-pages: the last event for which the geophone data and several photos are shown from the moment when the geophone data exceeded the defined thresholds; the station Laguna 513 with the measurements of the geophones and small images taken by the two cameras and an indication of the charge level of the battery; the station Pampa Shonquil with the plots of the measurements of precipitation, humidity, radiation, temperature, pressure sensor, and wind speed and an indication of the charge level of the battery; the cameras showing the recent photos in higher resolution; an SMS warning module with the possibility to send out text messages manually to the registered cell phones; and the alarm activation module; a data download sub-page with the possibility to download all archived dated for a selectable time range. The sub-page of the Pampa Shonquil station with the meteorological measurements is publically accessible, whereas the information on the other sub-pages is only available for registered persons from the involved authorities and the project team.

While the monitoring and warning system was being designed and implemented, an action plan was elaborated in collaboration with the main stakeholders (the members of the Local Emergency Operation Center (COEL), the Civil Defense Platform, selected government members, and the mayor). The plan defines all actions to be taken according to different warning levels, as well as contact details (names and phone numbers) of responsible persons and their deputies. The definition of this action plan had to take into account local, regional, and national laws, rules, and guidelines.

If the measurements of one geophone exceed a defined threshold, an SMS is sent out to all involved persons automatically, telling to immediately check the EWS data online or in the data center. Subsequent steps have then to be taken according to the predefined action plan and based on the data available (Fig. 16.4). In this EWS, no alarm is launched automatically by the system. The thresholds of the geophones, used to activate the warning levels, are defined on the basis of a continuous data analysis which is currently being carried out jointly by different Peruvian and Swiss Institutes.

16.3.4 Response Capability (Education and Preparedness)

It is important that potentially affected people adequately understand the goals and procedures of the EWS and share a common interest in these efforts. It is therefore crucial to inform and train the public and the involved authorities, both technically and socially. Each actor needed a different training, according to their responsibilities and capabilities. A process of institutionalization has been carried out for local authorities (Municipality of Carhuaz, Civil Defense) in order to increase the efficiency (clear and transparent rules and simplified procedures for the implementation of risk management). It is also worth mentioning the work with the population with whom information and awareness workshops have been carried out about the real state of the danger they face. Carey et al. (2012) noted that when the population does not participate in the processes of risk management, they see the results as an imposition, causing them to resist management actions even though these will help them to save their lives.

To embody and apply learned lessons, test alarms and evacuation simulations are very effective means to train both authorities involved in the EWS and civilians (Frey et al. 2014). Such simulations can be used not only to expose the population to an evacuation under near-realistic conditions but also to test the action plan and the decision process of the responsible authorities.

Civil response to the EWS is a key aspect when assessing its effectiveness and has been tested several times since its implementation, and emergency simulations are scheduled every year for the entire country. These practices are essential for

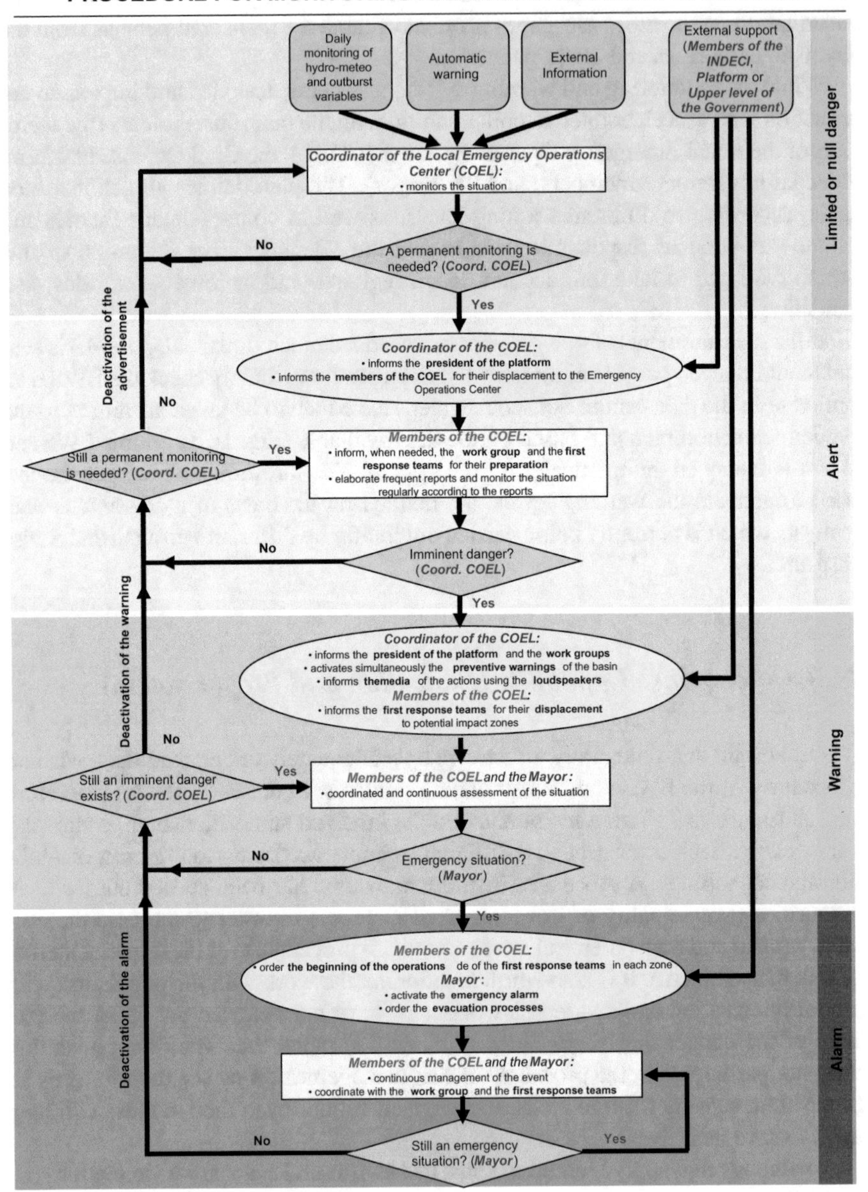

Fig. 16.4 Flowchart-type action plan for the different warning levels

improving authorities' knowledge of how well communities carry out the emergency procedures as well as to validate the feasibility of the evacuation routes derived from the hazard maps.

16.4 Lessons Learned and Perspectives

This is the first EWS of its type implemented in Peru related to risks of glacier lakes, and as such, it has been a source of important lessons (technical as well as institutional) which are important for the planning of subsequent systems.

During the progress of the project, positive aspects of the implementation of the EWS, main issues encountered as well as the working areas where supplementary efforts or new approaches are needed to accomplish the objectives were identified. Moreover, as the project "Glaciares 513" reached the end of its first phase, an external and independent evaluation of the project performance identified the strengths and weaknesses regarding the implementation and exploitation of the EWS according to the following five criteria: pertinence, efficacy, efficiency, impact, and sustainability.

The *main positive aspects* are listed below:

1. **An overall positive assessment of its implementation**, even more so considering that the specialized focus on adaptation and risk management related to receding glaciers could have presented limitations. The efficacy of the project has been demonstrated as it has had the support and involvement of the population and the local authorities in activities related to glacier risk reduction (through the design and implementation of the EWS).
2. **A good reception of the project and the EWS** in the public and private sectors, academic and practical area, at the national, regional, and local level. The workshops carried out by the leaders of the project had a positive impact on the main stakeholders involved in the GLOF's quandary in the region. Receptivity to the implementation of an EWS of the population has been crucial because without their participation, its design would not have been possible.
3. **A good response to test alarms and evacuation simulations** performed during the past years. Population, local authorities, and civil defense organizations assimilated the importance of established protocols and carried out them with good results, which shows that all stakeholders appreciate the utility of such a risk management tool.
4. **The installation of the technical part of the EWS** (sensors, stations, etc.) was executed with the support of local staff and was accomplished without major problems. Nevertheless, it is worth mentioning the lack of experience in Peru in maintenance and operation of such systems under high-mountain conditions, which had an effect on the correct operation of the proposed EWS.

Moreover, *problems encountered and improvements to be accomplished*, particularly those related to technical aspects of the EWS, are the following:

1. **Some communication difficulties** between the project partners and the final users of the system (i.e., the Municipality of Carhuaz). This represents an opportunity to reinforce the crosswise strategies into communications, gender, monitoring accountability, and transparency in the future.
2. **Key actors involvement** during the project activities. Several factors—such as not all the key actors being involved since the beginning of the project, or the

political changes in the local government of Carhuaz—resulted in a lack of comprehensive assessment of the social and political situation of the areas of intervention, and contributed to the delay in the process of implementation of certain activities and challenged the sustainability of the EWS.

3. **Recurrent technical problems in the EWS** causing a delay in its final operation. The proper operational functioning of the different components of the implemented EWS is highly challenging in such extreme conditions in high mountains. Several persistent technical problems (bad or null signal reception between the stations, failures due to lightning impacts, energy problems, etc.) repeatedly disturbed the operation of the system. This led to unexpected expenses, change of sensors and their disposition, and a lack of data for defining warning thresholds for the geophones signals. A proper diagnosis of the problems and an integral correction of the system scheme were then needed.
4. **The alarm module of the system (sirens and their corresponding infrastructure, permanent surveillance in the municipality, etc.) was not yet implemented**. This task was in the responsibility of the Municipality of Carhuaz. It is planned to install one or two long-range sirens to cover the entire area of the city of Carhuaz. In addition, the system has the possibility to send out predefined text messages to district leaders in parallel to the acoustic alarm. The populated places further up in the catchment are not included in the acoustic alarm concept at the current stage. However, they could be included in the described SMS service, given the coverage of the signal.
5. **Geophones' thresholds for the activation of the different warning levels were not yet defined**. At the end of the first phase of the project, the system was not yet fully operational (even if it has been tested on several occasions) because the thresholds for the geophone measurements described above that indicates when a warning level is activated were not established yet. In this test phase, three different thresholds were to be defined to make sure that smaller events got registered as well (lowermost threshold) and to have a direct indication of the size of the event, depending on only one, two, or even all three thresholds are overshot. For this, historical records of the geophones had to be taken into account, and data analysis had to be applied to them as well.
6. **Further involvement of other technical fields such as electronic engineering is required**. The multiple technical problems identified in the EWS, most of them caused by unexpected power source issues and deficiencies on the sensors and transmission devices, could have been solved or even avoided if an initial well-designed scheme had been elaborated which relates back to missing local experience. This could also lead to a more participative role of universities and/or research platforms on the conception and design of this (or other) EWS, benefiting both the technical aspects of the project and potential research initiatives.
7. **Toward a multipurpose project**. When assessing the scope of the EWS in the context of the "Glaciares 513" project, a broader perspective of adaptation to effects of climate change results in an integration of risk reduction and water resource management in the light of dwindling water resources from shrinking glaciers. Multipurpose projects have the objective to integrate both perspectives

and thus generate multiple benefits. For the "Glaciares 513" project, a set of complementary measures were considered and evaluated with different government institutions: the construction of an additional tunnel system through the bedrock dam 30 m lower than the existing ones, with the possibility to control the discharge and further reduce GLOF risk and create reservoir capacity to release water during the dry season; a monitoring of slope stability in the surrounding of the lake with spaceborne radar interferometry (InSAR); or further climate change adaptation measures supported by local communities.

The identification of strengths and deficiencies is of importance since this pilot experience is used as a major reference for similar EWS at the national scale in Peru. A second phase of the project, called "Glaciares+. Risk Management and the Productive Use of Water from Glaciers", started in November 2015 aiming at completing and providing continuity to the incomplete processes from first phase (especially the implementation of adaptation measures to the EWS) while support to further EWS in other catchments will be provided. In order to overcome the difficulties encountered during the "Glaciares 513" project, this second phase is conceived to strengthen public management related to the issue of GLOFs, stress the legal framework as well as the planning, implementation, and monitoring of this kind of project.

The appropriate identification of technical, social, or institutional aspects to be improved is going to entitle the main leaders in the design of more robust, efficient, and resilient projects related to glacier's risk management.

16.5 Conclusions

In the context of the project "Glaciares 513", an early warning system was designed and implemented at Laguna 513 of the Cordillera Blanca (Ancash, Peru). The main objective was to provide the potentially affected communities with an integrated tool to reduce GLOF-related risks, most importantly to avoid any loss of live. The presented EWS is a nonstructural measure, lowering the risk by reducing the damage potential through evacuation of the population in case of an outburst. It was elaborated according to international and national standards, and consisted in the four standard components of an EWS: (i) Risk Knowledge through the modeling of GLOF events and the elaboration of hazard maps, following international standards (cf. Schneider et al. 2014); (ii) Monitoring and Warning Service, with the physical components of the system, that is the ensemble of sensors, stations, and broadcasters that detect and transmit hazard parameters to generate warnings in a timely fashion; (iii) Dissemination and Communication, through the establishment of an action plan and protocols for the management of the information generated during the detection of potentially hazardous events; (iv) Response Capability of affected population, carried out through educational programs and the performance of evacuation simulations.

In general, the EWS had a positive impact on affected communities and a good reception by local authorities as well as social services and civil defense; the assimilation of the system by the main stakeholders was satisfying.

However, this pioneer system faces some challenges, especially when it comes to the proper functioning of electronic components (such as sensors, signal broadcasters, and batteries). The lack of continuous base data makes the definition of the geophone's warning alert thresholds yet difficult. More implication of other technical fields such as electronics is required to overcome these issues. Moreover, the involvement of key actors and the communication with them must be redefined.

By identifying these problems and weaknesses, a second phase was outlined, in which a multipurpose approach would enlarge the scope of the project and optimize the outputs. The goal of this second phase of the project ("Glaciares+, Risk Management and the Productive Use of Water from Glaciers") is to complete and provide continuity to the various unfinished processes from first phase, especially the implementation of adaptation measures to the EWS. As the main outcome of this phase, the system needs to achieve full sustainability to guarantee that the processes and activities necessary for monitoring and reducing glacier-related risks are maintained over time, generating constant and permanent results.

NOTE This article was written in May 2016; at this time, the EWS was intact and fully functional. In November 2016, a group of people accessed the monitoring station at Laguna 513 and destroyed its main components (Fraser 2017), rendering it completely useless. However, the authors consider that the lessons learned until then are still valid and of interest for the design, implementation, and improvement of such systems.

Acknowledgements The studies and works presented in this paper have been conducted under and with support of the "Proyecto Glaciares", funded by the Swiss Agency for Development and Cooperation (SDC), executed by CARE Peru, the University of Zurich, CREALP, Meteodat, Swiss Federal Institute of Technology Lausanne (EPFL), and local partners such as the Unidad de Glaciología y Recursos Hídricos, Autoridad Nacional de Agua (UGRH, ANA), and others such as the Ministerio de Ambiente, Peru (MINAM) and the National Park Service (SERNANP). We acknowledge the collaboration of several further colleagues of these institutions.

References

ANA (Autoridad Nacional del Agua), Inventario Nacional de Glaciares y Lagunas (Huaraz, 2014)

Bulmer, M. H., & Farquhar, T. (2010). Design and installation of a Prototype Geohazard Monitoring System near Machu Picchu, Peru. *Natural Hazards and Earth System Sciences, 10,* 2031–2038. https://doi.org/10.5194/nhess-10-2031-2010.

Carey, M., Huggel, C., Bury, J., Portocarrero, C., & Haeberli, W. (2012). An integrated socio-environmental framework for climate change adaptation and glacier hazard management: Lessons from Lake 513, Cordillera Blanca. *Peru. Climatic Change, 112*(3–4), 733–767. https://doi.org/10.1007/s10584-011-0249-8.

Christen, M., Kowalski, J., & Bartelt, P. (2010). RAMMS: Numerical simulation of dense snow avalanches in three-dimensional terrain. *Cold Regions Science and Technology, 63*(1–2), 1–14. https://doi.org/10.1016/j.coldregions.2010.04.005.

Fraser, B. (2017). *Learning from a flood-alarm system's fate*. GlacierHub. http://glacierhub.org/2017/05/31/learning-from-a-flood-alarm-systems-fate. Accessed October 24, 2017.

Frey, H., García-Hernández, J., Huggel, C, Schneider, D., Rohrer, M., Gonzales Alfaro, C., Muñoz Asmat, R., Price Rios, K., Meza Román, L., Cochachin Rapre, A., & Masias Chacon, P. (2014). An early warning system for Lake Outburst floods of the Laguna 513, Cordillera Blanca, Peru. In *Paper presented at the Analysis and Management of Changing Risks for Natural Hazards International conference*, Padua, Italy, November 18–19, 2014.

Huggel, C., Khabarov, N., Obersteiner, M., & Ramírez, J. M. (2010). Implementation and integrated numerical modeling of a landslide early warning system: A pilot study in Colombia. *Natural Hazards, 52,* 501–518. https://doi.org/10.1007/s11069-009-9393-0.

IBER. (2010). Two-dimensional modeling of free surface shallow water flow, Hydraulic reference manual, IBER v1.0.

ISDR (UN International Strategy for Disaster Risk Reduction), Developing early warning systems: a checklist. In *Paper presented at the 3rd International Conference on Early Warning*, Bonn, Germany, March 27–29, 2006

Kääb A., Reynolds J. M., & Haeberli W. (2005). Glacier and Permafrost Hazards in High Mountains. In Huber U. M., Bugmann H. K. M., & Reasoner M. A. (Eds.), *Global change and mountain regions. Advances in global change research* (Vol. 23, pp. 225–234). Dordrecht: Springer.

Mark, B., & Seltzer, G. (2005). Evaluation of recent glacier recession in the Cordillera Blanca, Peru (AD 1962–1999): Spatial distribution of mass loss and climatic forcing. *Quaternary Science Reviews, 24,* 2265–2280. https://doi.org/10.1016/j.quascirev.2005.01.003.

Muñoz, R., Gonzáles, C., Price, K., Frey, H., Huggel, C., Cochachin, A., García, J., & Mesa, L. (2015). Managing glacier related risks in the Chucchún Catchment, Cordillera Blanca, Peru. Abstract presented at the EGU General Assembly 2015, Geophysical Research Abstracts, Vol. 17, EGU2015-13131. Vienna, Austria, 12–17 April.

Racoviteanu, A., Arnaud, Y., Williams, M., & Ordoñez, J. (2008). Decadal changes in glacier parameters in the Cordillera Blanca, Peru, derived from remote sensing. *Journal of Glaciology, 54*(186), 499–509. https://doi.org/10.3189/002214308785836922.

Raetzo, H., Lateltin, O., Bollinger, D., & Tripet, J. (2002). Hazard assessment in Switzerland—Codes of Practice for mass movements. Bulletin of Engineering Geology and the Environment, pp. 263–268. https://doi.org/10.1007/s10064-002-0163-4.

Schneider, D., Huggel, C., Cochachin, A., Guillén, S., & García, J. (2014). Mapping hazards from glacier lake outburst floods based on modeling of process cascades at Lake 513, Carhuaz, Peru. *Advances in Geosciences, 35,* 145–155. https://doi.org/10.5194/adgeo-35-145-2014.

Vuille, M., Francou, B., Wagnon, P., Juen, I., Kaser, G., Mark, B., et al. (2008). Climate change and tropical Andean glaciers: past, present and future. *Earth-Science Reviews, 89,* 79–96. https://doi.org/10.1016/j.earscirev.2008.04.002.

BY

Chapter 17
Enhancing Frontline Resilience: Transborder Community-Based Flood Early Warning System in India and Nepal

Yeeshu Shukla and Bhanu Mall

17.1 Introduction

Terai[1] belt in India and Nepal is considered one of the most economically backward areas due to various historical and socio-economic factors. The area is also affected by recurring floods. The ecological settings through which flooding takes place in Terai are also diverse. For years, floods have ravaged the lives of the inhabitants of this area with unrelenting regularity every alternate year and many a times for subsequent years. The current trends show frequent greater flood problems in the area than ever before thus resulting in the reversal of any gains on the developmental front (Singh 2009). The plights of the people are aggravated by regularly affected crops, health and employment due to regular floods in this region.

There have been significant flood events since 1971 leading to loss of lives and livelihoods on a regular basis. In July of 1993, Nepal experienced a devastating flood in the Terai region, which took the life of 1289 people and affected 575,000 people. In 1998, flood-affected again about half million inhabitants and caused the total loss of about 2 billion Nepal rupees (approximately $US29 million). The Saptakoshi River floods and landslides in 2008 affected 28 districts of the country and impacted more than 300,000 people (Global Risk Identification Program). Similarly, in the Indian part of Terai, floods have affected caused considerable damage to the poor and marginalized. In 1998, 127 people lost their lives and more than 90,000 households

[1] Terai, also spelled Tarai, region of northern India and southern Nepal running parallel to the lower ranges of the Himalayas.

Y. Shukla (✉)
Christian Aid Organization, London, UK
e-mail: yshukla@christian-aid.org

B. Mall
Poorvanchal Gramin Vikas Sansthan (PGVS) Organization, Lucknow, India
e-mail: drbhanu53@gmail.com

S. Hostettler et al. (eds.), *Technologies for Development*,
https://doi.org/10.1007/978-3-319-91068-0_17

were affected in the area. In the year 2008, some 170 people died and more than 2.5 million were affected with 165,000 people getting displaced (Singh 2009).

17.1.1 Nuances of Vulnerability

The lessons from previous interventions on early warning in the region have shown that there are some specific factors that contribute to the increased vulnerability and risk at the frontlines. These have been documented during the course of action by Christian Aid and its partners. Some of them are as follows.

Weak physical and natural resources: Weak physical and natural assets define communities vulnerable to flood. There are instances where retaining walls break, overgrazed lands subsides and dangerous mudslides, bridges collapse and crops wash away. Often communities with the weakest physical assets are the same communities that are located closest to the source of disaster—on the banks of river, for example. Since primarily poor live in such places with no means of ensuring any protection, these places often get exposed to the numerous disasters.

Degraded natural environment: Degraded lands are more susceptible to damage, during natural events. Healthy, biodiverse forests or river banks/coasts with vegetation can absorb water to some extent. Land abundant grasses and vegetation also soak up excess water. Overgrazing and deforestation have led to destructive landslides, and crops unprotected by bunds and other conservation measures wash away.

Inadequate access to support structures: Communities located in the most fragile environments are the same communities that lack the social and political assets needed to attract important government resources—better roads, strong bridges, sturdy tanks and shelters, or relief supplies. Poor access to government and non-government support impedes a community's ability to withstand disaster.

Lack of disaster preparedness: Communities in flood-prone areas already have coping strategies to address flood, but they are often insufficient in the face of major flood. Good planning (and ensuing implementation) is, for vulnerable people, the best defence against disaster. This is especially true for marginalized communities living in vulnerable locations—sick, handicapped or elderly people, small children, expectant mothers and resource-poor families who have been pushed to dangerous ground or who lack the means to escape or survive.

Inadequate mechanism for timely information flow: Most of the communities within flood-prone areas are devoid of any form or reliable information which can be used to prepare for upcoming disasters. Often they come to know about floods when it strikes their habitat and they are affected by it, by then it is too late to respond. It is not that there is lack of information about the onset of disaster. However, the issue is of its dissemination to communities vulnerable to it. Based on field assessment, there is clear evidence of information sharing between authorities on Nepal (Department of Hydro-Meteorology) and India (irrigation department); most of these information remain within the administrative setup and seldom reaches the desired recipient. Therefore, it is desirable to have early warning mechanism which is based in community and can be managed locally without much technology. It is vital that sustainability is imbibed into the core of its principle.

17.1.2 Why Disaster Risk Reduction?

The above lessons have been derived and documented by working with the communities in the region which brings out a clear need for a robust disaster risk reduction mechanism having a potential to minimize disaster risk for the communities living at the frontlines. The Sendai Framework of Disaster Risk Reduction (SFDRR)[2] promotes such a people-centred approach to early warning as one of the priority areas for communities to avoid risks. The objective of people-centred early warning systems is to empower individuals and communities threatened by hazards to act in sufficient time and in an appropriate manner so as to reduce the possibility of personal injury, loss of life, damage to property and the environment and loss of livelihoods

The underpinning assumption is that this requires strong community participation on the ground and enabling environment for greater cooperation across various stakeholders including the government authorities in India and Nepal across levels on knowledge and information sharing. As it is known that the resilience agenda could also move to become a more integral part of ongoing local development processes, so it would be appropriate to look into the above processes and policies in the mainstream development planning.

17.2 The Initiative

In order to address some of the above-mentioned concerns, Christian Aid with its local partner Poorvanchal Gramin Vikas Sansthan (PGVS) initiated a pilot on flood early warning systems in the Terai region of India and Nepal on the Karnali (Nepal)–Ghaghra (India) river basin. The initiative is provided technical assistance by Practical Action and financial assistance from Cord Aid. This involved accessing data available with relevant government departments, dissemination and communication and collaborative arrangements with the private sector for IT-based solutions. The project in the initial pilot phase has achieved some tangible results in terms of institutionalizing the effective community preparedness mechanism utilizing and demystifying the technical knowledge available from both Nepal and India. The project focuses on following broad objectives:

[2]The Sendai Framework for Disaster Risk Reduction 2015–2030 (Sendai Framework) is the first major agreement of the post-2015 development agenda, with seven targets and four priorities for action.

1. To establish transborder community-based early warning system using Information and Community Technology along the Karnali River in Nepal and Ghaghra River in India with the focus on strengthening frontline resilience.
2. To build the capacity of local community-based institutions on effective flood preparedness with clear linkages between upstream and downstream.
3. To develop formal and informal communication channels to access early information on water levels from upstream in Nepal.
4. To strongly advocate for having effective community-based Early Warning System (EWS) across flood-prone Terai in India and Nepal (Fig. 17.1).

17.2.1 Information and Communication Technology (ICT) for Flood Resilience

The EWS system being implemented in the Terai region of India and Nepal offers huge potential for bringing technology and community action together by demystifying the concept of technology-centred approaches involving people who are expected to be affected by the disasters. The system involves following key elements and processes.

17.2.2 Monitoring and Warning

The system begins with facilitating capture of information for monitoring and dissemination of warning with the incorporation of ICT solutions suited to emergency situations. The water levels are monitored in the upstream at Chisapani in Nepal and cross checked with the information being generated by Department of Hydro-Meteorology, Government of Nepal. The same information is channelized to the India side. Christian Aid partner-PGVS with the help of Gram Panchayat[3] (local elected body) and community groups have established electronic display machine in district Bahraich and Gonda in India. The display machine shows water level in Karnali River at Chisapani in Nepal. The digital river monitoring system receives the readings directly from Department of Hydro-Meteorology (DHM) through Internet, and it reflects information in real time. This provides a fair idea of the lead time that the communities in the downstream have depending on the water levels.

The system is also equipped with alert alarm system which gets activated based on the water levels in Nepal. The alert is based on the threshold limits based on past experiences of water levels and floods in the region. The recorded thresholds are shown below.

[3] A Gram Panchayat is the cornerstone of a local self-government organization in India of the Panchayati raj system at the village or small town level, and has a Sarpanch as its elected head.

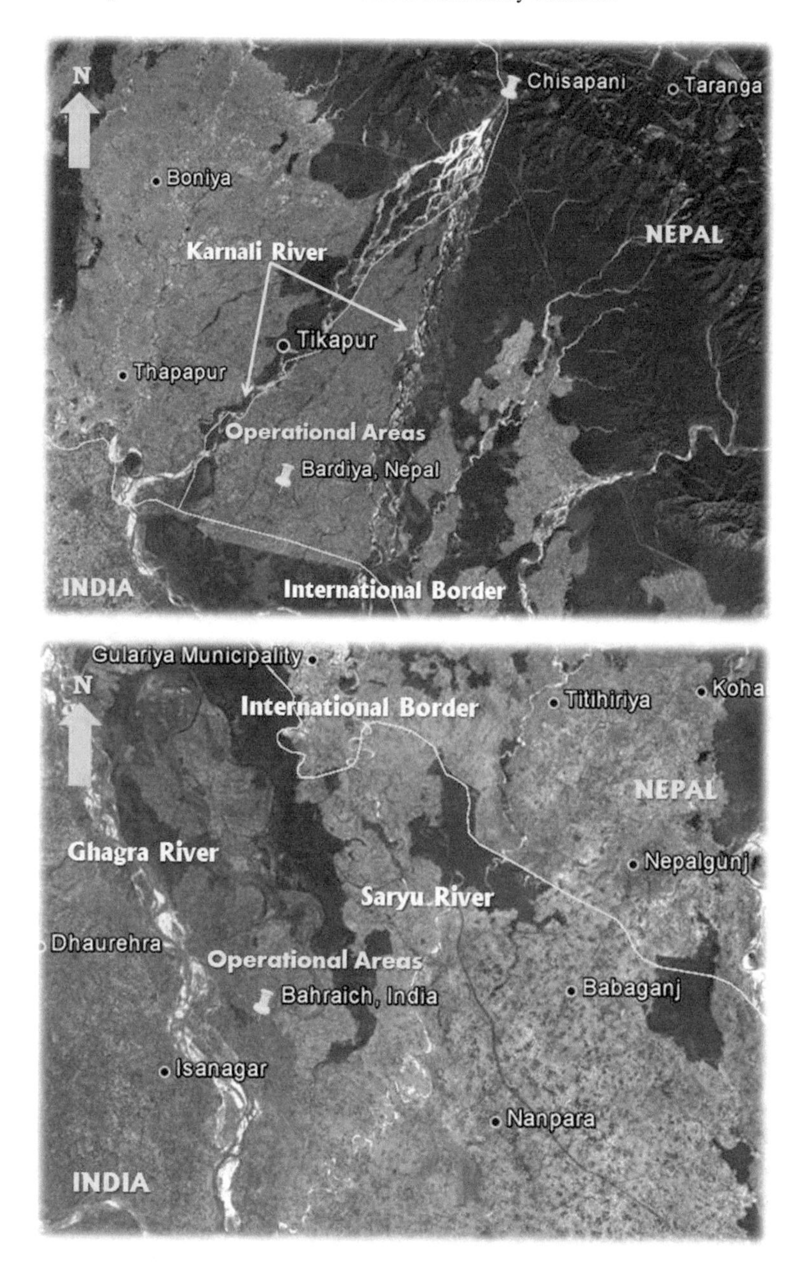

Fig. 17.1 Operational areas in India and Nepal. *Source* Google Earth

Figure 17.2 provides an example of water levels in Nepal and India and the respective alert and warning levels. The data shown in the alert is not taken from Mean Sea Level (MSL) but the level of water recorded at the points where rain

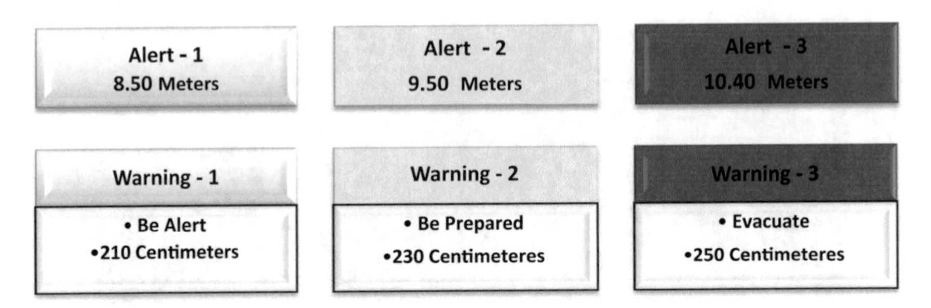

Fig. 17.2 Example of warning and alert levels based on the thresholds recorded from the past experiences. *Source* PGVS

gauges have been established in Nepal and India. The alert levels are fairly specific to a particular community and warnings are issued for a particular community based on the water levels compared to pre-monsoon levels. Hence, the system takes into account a locale-specific vulnerability, and exposure to rising waters and provides a context-specific warning.

17.2.3 Generating Data from the Frontlines

The project has been instrumental in the installation of river gauges at strategic locations close to the villages which get affected due to the floods in the region. The water level is periodically monitored at these locations along with the data generated by DHM. The information is then corroborated for issue of alerts from the district hub. It is important to mention that this information is generated by the communities trained as task forces under the initiative from the project villages.

17.2.4 Dissemination of Alerts and Warnings

In addition to the electronic display boards in the communities which could lose connectivity during floods, information is shared with voice and text messages. The mobile numbers of key personnel both in the communities and district and state level functionaries are registered for dissemination of information.

The contacts are arranged in groups based on the project area and can be subdivided as well. This gives a quick access during emergencies. The information is directly sent to all the mobiles registered in the group and information can be given in local language as well (Hindi). This is handled by authorized personnel of communication department of the PGVS or anyone assigned by the authority. Similarly, voice messages are sent with demystified information to the people expected to be affected by the rising waters.

17.2.5 Feedback Mechanism

The feedback promoted by the system uses ICT mechanism of helpline number. The service is known as super receptionist service and is based on Interactive Voice Service (IVR). The number is Indian and local charges apply which is based on network. During emergency situations, the lines are opened which promotes two-way communication and encourages people to feedback into the system.

Super Receptionist IVR combines technology with the available tools. The listing and grouping as an outgoing service is also present along with 24 h incoming service. ICT technology improvises inductive use of telecom services and appropriate for the emergency response. As far as monitoring goes, no external resource needs to be adapted as internal mechanism of recording calls and download the report facility is available (PGVS 2016) (Fig. 17.3).

17.3 System in a Real World Situation (Case from Village Somai Gauri)

17.3.1 Background

Village Somai Gauri is situated in Mihinpurwa Block in district Bahraich is around 97 km from district administrative office. The Palihapurawa hamlet situated under this Gram Panchayat is situated in north-east direction. The Log and Lat of this village is 27 58 37 N and 81 15 34 East. The total population of this village is 765 (Male 390 and Female 375) with most of them belonging to backward caste. The Gram Panchayat is located very close to the Ghaghra River that originates from Nepal and popularly known as Karnali within Nepal. Floods are a regular phenomenon in this Gram Panchayat which causes widespread devastation to the lives and livelihoods.

The river is constantly changing course and the distance of the River from village, which used to 6 km some 6–8 years ago, now reduced to less than 1 km. Growing paddy crop is impossible in these areas due to the heavy flood. Only sugarcane and corn crops are main agricultural products of these areas which also get affected at times. The early warning system was initiated in the village on 1 April 2012 for enhancing the flood resilience of the communities through better coordination and linkages between upstream and downstream communities.

17.3.2 Local Action by People

The initiative began with engaging people in their self-analysis of existing vulnerability and capacity. A systematic approach called Participatory Vulnerability and Capacity Assessment (PVCA) were carried out and Task forces at village level and Village Disaster Management Committee at Gram Panchayat level were formed and

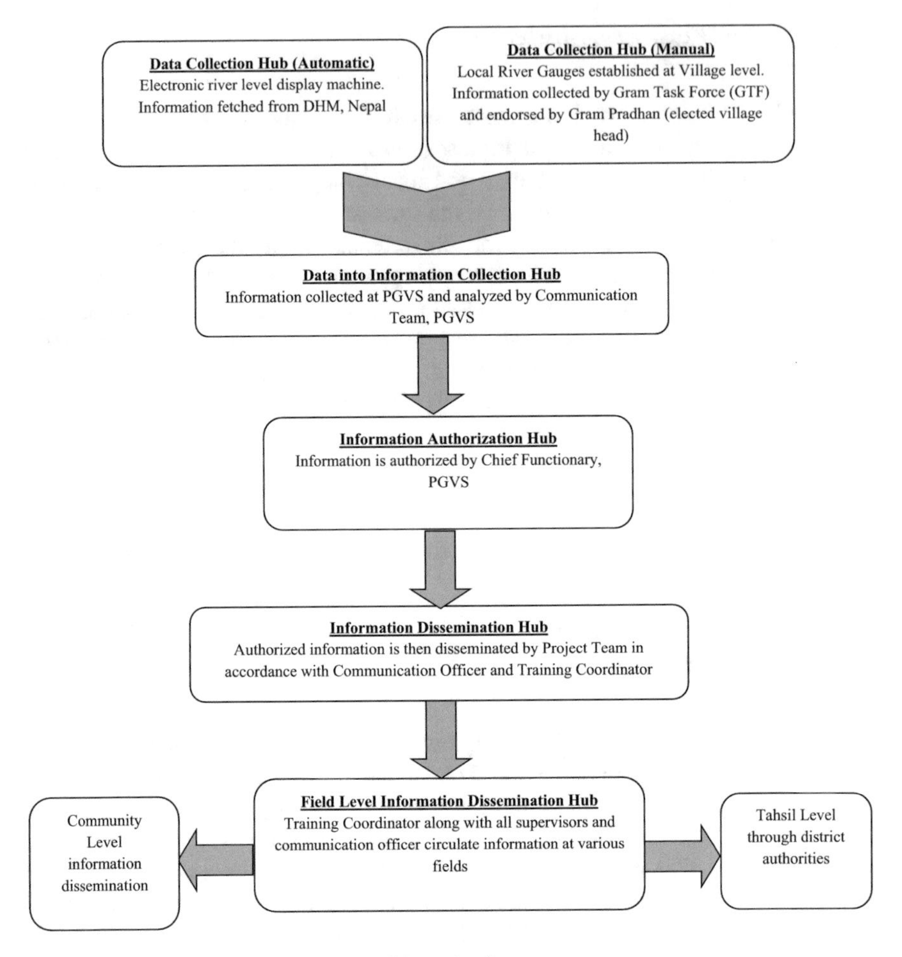

Fig. 17.3 Flow chart of the transborder ICT mechanism

linked to the established community-based early warning centres under the leadership of Gram Panchayat (local elected body). The river monitoring display board has been established in Palihapurawa, district Bahraich and Pure Baloo, district Gonda in coordination of Department of Hydro-Meteorology, Nepal, and River gauges are also established in villages by Gram Task Force (GTF) under the supervision of Gram Panchayat. The Gram Panchayat and GTF have taken responsibility for maintenance and safety of the equipment.

The GTF has divided into three groups like Early Warning and Communication Group, Search and Rescue Group and First Aid and Shelter Management Group, consisting of 15 members of Gram Task force in each village. Gram Task force has been trained on search and rescue, first aid and shelter management. The Gram Task force has all the maps which were developed during PVCA exercise including important contact numbers related to Disaster and Development Department, information on engineers associated with local dams and their contact number, etc., the life-saving equipment like life jackets, floating rings, ropes with carabineers, and siren's has been kept in village community-based early warning centre. The Gram Task forces have been involved in conducting mock drill in their village before the monsoons.

17.3.3 Floods of 2013

In the year 2013 when the area faced severe floods, villagers from Somai Gauri were able to evacuate in time with their personal belongings to safer places. They utilized the information generated from the EWS and the training that they received from the project team.

When the water levels started rising, Early Warning and Coordination Group member discussed with all Gram Task forces including village head about the changing river trends and started taking data from river monitoring system four times in a day (6 AM, 10 AM, 2 PM and 6 PM each day) and shared the status with the president of the Gram Task force. The same was also communicated with the village head, PGVS team and sub-divisional office, Nanpara through mobile phone every day. PGVS early warning team then shared this information through text message and voice messages with multiple stakeholders. The Gram Pradhan (village leader) puts his signature on river monitoring register at 6 PM every day in the evening. The task force based on the warnings and rising level of water placed a one corner white flag at early warning centre and peals the siren for 30 s as alert situation to the villages.

The Gram Task Force member simultaneously disseminated flood information to villagers in the nearby villages by the existing communication channel. After this first alert, people started organizing their domestic vital materials, important documents, livestocks, ornaments and food materials.

The water level of Ghaghra River at Chisapani, Nepal crossed its warning level and reached 10.24 m on 17 June 2013 at 6 PM evening and reached 1.80 m on the same day reflected in the village River gauge located close to the village. A rapid call to all GTF members was given and Early Warning and Coordination Group was activated. Now village head asked GTF members to place the two corner white flag and peal the siren two times for 30 s. The GTF members disseminated the flood information to the village heads of other nearby villages and communities started prepositioning of food and other essential. The flood waters entered agriculture land on 18 June 2013 at 5 am, and slowly water entered the village. Mr. Ram Pal took quick action and informed village head about the situation through mobile and village head quickly asked the task force to place a three corner white flag at EWS centre and peal the siren continuously in the village.

17.3.4 Benefits

By these efforts and quick responses of the Gram Pradhan and GTF members, communities could manage to evacuate safely from the village and reached Jalimnagar Bridge with all required materials and established their shelter during flood period. The flood water stayed in village for 2 days at the height of about 3–4 ft and started reducing from 4 PM on 20 June 2013. There was no loss of lives due to the advance information and improved coordination among community and local institutions.

17.4 Conclusion

Information on climate change is building a new perception of disasters as of our own making. The increase in floods, storms, droughts and other hazards expected to arise from the accumulation of greenhouse gases in the atmosphere as a result of industrialization and deforestation is clearly not natural (UNISDR 2008). Most of the literature and scientific consensus now confirms that no matter whatever we do to reduce the greenhouse gas emissions, the human-induced changes in climate is inevitable. So this brings forth the need for reduction in emissions and most importantly being prepared for changes holds the key.

Effective preparedness and early communication to these growing changes will require strategies, institutions and structures that are robust under uncertainty. Moreover, as communities are directly getting affected, their own coping mechanisms, capacities and willingness to adapt newer technologies are of extreme relevance and need to be studied in detail. The traditional knowledge used as coping mechanisms as also the new mechanisms developed over the years provide greater resilience to the communities in times of growing climate threats. At the same time, it is equally important to understand the existing policy regime in the context of lifeline communication systems like early warning mechanism in transboundary river systems with a dynamic fluvial landscape and other overriding geopolitical concerns.

The robust EWS systems also present an opportunity to look into the feasibility of ICT technology having strong human/community linkages for scaling up in other regions with similar vulnerability, hazards and risks with a focus on transborder water resources management for disaster risk reduction in times of changing climate. However, there are several challenges in implementing EWS systems in a transborder context and some of the actions required to strengthen the EWS systems could be as follows.

17.4.1 Generating Buy in for EWS

To scale up the EWS in the multi-hazard zones, it is fairly important to have a favourable environment which includes having a clear buy in with the government systems at various levels, non-government organizations and communities at the frontlines. This could be done with piloting such approaches and documenting the evidence of the benefits of the approach for at-risk communities. The evidence generated needs to be effectively communicated to the government and other actors for taking this forward and making it a priority action for minimizing the loss of lives and livelihoods.

17.4.2 Accelerating Regional Information Sharing

One of the major bottlenecks in streamlining and implementing such systems is related to lack of effective regional information sharing mechanism between respective governments. India and Nepal have a Joint Committee on Water Resources (JCWR) and the last meeting of the committee took place in January 2013. This is a good mechanism for streamlining regional cooperation. However, the frequency of the exchanges needs to be enhanced and the discussion needs to take specific hazards and vulnerabilities that the region faces due to the floods. Similarly, civil society initiatives on both sides need to be explored involving people from all walks of life. Christian Aid and its partner are trying to create such a mechanism between the civil society groups of both the countries.

17.4.3 Participation, Social Cohesion and Active Involvement

Participation and active involvement of the at-risk communities is the backbone of the high user interface ICT systems. Another important aspect is related to co-creating with potential participants who would help innovate and troubleshoot bottlenecks on a regular basis. This could also lead to development of multiple co-creation teams having insights on specific physical, social and economic context.

17.4.4 Capacity Building as Ongoing Activity

Actions need to be substantiated with continual capacity building inputs. Often, it is seen that capacity building is seen as one-time activity. The content should be based on the local knowledge base in line with the technical infrastructure. Efforts should be focused on simplifying the use of technical aspects of the system which could be used by a large number of population. This is critical for scaling up the innovation to other areas and increase uptake. Gaps need to be identified during the implementation phase and specific modules need to be developed to address these gaps.

Acknowledgements We would like to offer our sincere thanks to Mr. Krishna Tripathi and Mr. Shivanshish Snha (both PGVS staff) for their inputs on the paper.

References

24–25 January 2013, Kathmandu, *Minutes of the Seventh Meeting of Nepal-India Joint Committee on Water Resources (JCWR).*

Ahmed, S., & Dixit, A. (2007). Working with the winds of change: toward strategies for responding to the risks associated with climate change and other hazards, IDRC and DFID

Asia-Pacific Input Document for the Post-2015 Framework for Disaster Risk Reduction (HFA2), 2014

Global Risk Identification Programme, Historical Disaster Profile of Nepal (based on NSET disaster observatory database).

Hyogo Framework of Action-2005-15

IUCN, & IISD. (2008). *Livelihoods and climate change: combining disaster risk reduction, natural resource management and climate change adaptation in a new approach to the reduction of vulnerability and poverty*. International Institute for Sustainable Development

Mercy corps and Practical Action. (2010). *Establishing community based early warning system, Practioner's handbook*.

Sendai Framework of Disaster Risk Reduction 2015-30

Singh, G. (2009). *The natural disasters in the Eastern Uttar Pradesh*, India

Summary report by IPCC, 2007, 2014

United Nations International Strategy for Disaster Reduction Secretariat, Geneva, Links between Disaster Risk Reduction, Development and Climate Change, 2008

Venton, P. (lead author), & La Trobe, S. (2008, July). Linking climate change adaptation and disaster risk reduction, Tearfund.

Chapter 18
Bridging the Information Gap: Mapping Data Sets on Information Needs in the Preparedness and Response Phase

Marc Jan Christiaan van den Homberg, Robert Monné and Marco René Spruit

18.1 Introduction

The poor face different levels of impact when exposed to natural hazards than the nonpoor (Hallegatte et al. 2016). In addition, data on the risks poor and vulnerable face and the actual impact they experience is often lacking. The digital divide plays an important role in this as developing countries and within these developing countries, the poor and vulnerable have less access to digital technologies (Leidig and Teeuw 2015). Apart from technological reasons, there are also political and social reasons. For example, the poor often get their income through the informal economy, and their land and resource rights are usually not registered.

This lack of data affects the actions of different actors in the preparedness and response phase. Three main actors can be distinguished in the response and preparedness phase (van den Homberg and Neef 2015). The affected community are the people directly and indirectly adversely affected and in need of urgent (humanitarian) assistance. The responding community consists of local or outside community members that support in relief or recovery but are not trained in crisis response. The responding professionals are part of the professional community in the field of disaster management, such as national and local governments, NGOs, and national crisis coordination centers. All these three groups have to decide based on the data and information available to them—in the period just before the disaster hits—which

M. J. C. van den Homberg (✉)
Cordaid and TNO, The Hague, Netherlands
e-mail: mvandenhomberg@redcross.nl

R. Monné · M. R. Spruit
Utrecht University, Utrecht, Netherlands
e-mail: robert.monne@ortec.com

M. R. Spruit
e-mail: m.r.spruit@uu.nl

S. Hostettler et al. (eds.), *Technologies for Development*,
https://doi.org/10.1007/978-3-319-91068-0_18

early actions are the best to prevent loss of lives and to protect livelihoods and—after the disaster has hit—which response actions to take.

Decision-making in the preparedness and response phase should be based on factual data about the needs on the ground, but is in reality a highly political process. For example, damage and needs assessments are competitive and difficult processes as agencies continue to see information on the impact of a disaster as something to be "owned" in order to leverage resources and influence rather than as something to be shared (Walton-Ellery and Rashid 2012). Responders also face "high levels of uncertainty, extreme stress with significant consequences of actions, compressed timelines and significant lack of information available initially followed by extreme information overload" (Preece et al. 2013). In the case of Typhoon Haiyan, for example, those responding considered the multitude of different information sources and formats generally as an information overload (Comes et al. 2013). Responders will have different degrees of data literacy and cognitive abilities to deal with these circumstances and to make sense of the data and information that is available and accessible to them. The degree of collective sensemaking that is reached among responders is another important factor in decision-making between and within organizations (Wolbers and Boersma 2013). Important in sensemaking is having reliable, accurate, and timely data. Getting such data right after a sudden onset disaster is challenging given the chaotic and disrupted situation. In the case of floods in Bangladesh, the working group on Disaster Emergency Response (part of the Local Consultative Group that coordinates between the Government of Bangladesh and the different development partners) established a Joint Needs Assessment project in 2014. The Joint Needs Assessment project started a large survey to collect data in the field given the absence of consistent, comparable information across the affected area, whereby they had to trade-off between time, quality, granularity, and available resources (Wahed et al. 2014). Harmonizing and coordinating the different assessments organizations are doing is a difficult task and heterogeneity issues in the data sets that come out of the assessments are most commonly unavoidable. Given these data-related challenges, it is evident that responders face information gaps. Whether it is not having enough information at the very onset of a disaster or whether it is having too much information later in the disaster; in both cases, their information needs are not adequately covered. These gaps will be more articulate in developing countries—often data poor and low tech—than in developed countries.

18.2 Research Questions and Methodology

Three research questions were defined: (1) What are the information needs of disaster responders so that they can take appropriate decisions? What are the associated timing constraints? (2) What are available and relevant data sources and when do they become available? (3) How do these data sources currently meet the information requirements? We conducted a case study to address these questions (Monné 2016). The case study was part of a community-managed disaster risk reduction (DRR) program that included setting up an innovative last mile early warning system using voice SMS and developing an app and dashboard to enable two-way information exchange between affected communities and responders. The pilot areas are riverine islands in northwest Bangladesh, so-called char-islands, which are part of the densely populated floodplains where many poor and vulnerable people live. The focus was on the most recent and severe river flood of the last years, namely, the floods of 2014 that affected almost two million poor and vulnerable people living in nine districts in northwest Bangladesh (Wahed et al. 2014). About 1 year after these floods, we performed 13 oral history semi-structured interviews of which 11 in Dhaka (national NGOs (active in the JNA consortium) and Department of Disaster Management) and two in Sirajganj (one with a farmer and fisherman, and one with the director and his two co-directors of the local NGO, MMS). We held one focus group discussion with seven disaster responders of MMS, one focus group with 15 people living on the chars (imam, teachers, entrepreneurs, part of the volunteer disaster management committees), and one focus group with 13 local government officials [Upazila and Union Disaster Management Committee, civil defense organization (Ansar VDP)]. So, in total, we got input from 51 people. We arranged the first batch of interviewees based on our existing network and such that we would have a representative cross section. Subsequently, we used a snowballing approach to grow our sample considering the availability of respondents and useful references. Although focal point in these sessions was the flooding of 2014, we did allow interviewees also to draw from their earlier or more recent disaster management experiences. All interviews were transcribed. The focus group discussions were done with an interpreter, usually at an open noisy marketplace, and could not be literally transcribed. Instead, we used the notes taken. All interviews and notes were subsequently labeled using NVIVO 10 for Windows and coded based on three themes, i.e., Activity, Decision, and Information Need. We used inductive coding to have subthemes emerge from the data. For each of these themes, clustering was done based on experience emerging from the familiarization phase, domain knowledge, and literature study. In addition, we asked the interviewees to validate our transcribed interviews. We asked two domain experts to validate and expand on the list of needs. We also used the lists of Activities and Decisions to identify possible discrepancies. For the second research question, we used, in addition to the interviews, Internet search and literature study. In that way, we could make an inventory of the data sets that were available during the flooding of 2014. For the last research question, we singled out all the indicators per data file and manually determined the match with a subtheme information need. We scored the match as Yes, No, or Partly. Afterwards, we used constrained COUNT formulae to calculate the coverage per disaster data source of the subtheme information needs. We used approximately the phases as defined in the Multi-Sector Initial Rapid Assessment (MIRA) (MIRA 2015) to label both the data sets as well as the information needs. The phases consisted of before (1), the first 72 h (2), the first 2 weeks (3), and the first 2 months (4). Table 18.1 gives an example for three data sets and information needs. Data B covers 33% of the information needs if no time constraints are considered. With time constraints, none of the information needs are met, since the information was needed already in phase 1 but came only available in phase 4.

Table 18.1 Mapping data sets on information needs

	Timing	Data A	Data B	Data C
Timing		1	4	3
Information need A	2	Yes	No	Partly
Information need B	2	Yes	Yes	Yes
Information need C	2	Yes	No	No

18.3 Results

(a) **Information Needs**
A small group of interviewees, especially at the local level, had difficulties expressing their information needs and identifying the type of decisions they had to take when directly asked for it. However, when interviewees where asked to describe their role in the flooding of 2014, it was possible for us to derive these. Information needs varied as well from one responder to the other, which could usually be attributed to differences in the organization they were working for, their specific expertise, and level of education. Table 18.2 summarizes the needs as emerged from the coding and clustering of the transcribed interviews in normal text. The list is not exhaustive given our limited sample size. In the italic text, we have added the needs that two domain experts contributed. We decided not to aggregate the information needs to a too large extent, given that we want to map the information needs to the information in the available data sets, but also to reflect the needs as they were expressed. We defined seven clusters for in total 71 information needs. We have put in Table 18.2 on the left clusters that relate to the Crisis Impact and on the right, those that relate to the operational environment, in line with the MIRA Analytical Framework (MIRA 2015). The cluster Damage and needs scored highest in terms of amount of times mentioned in all interviews and in terms of in how many interviews it was mentioned. This cluster of information needs matched also with what the interviewees mentioned as the most difficult decisions for them to take, i.e., determining which beneficiaries to support where and with what kind of support. Next comes the need for information on Coordination, especially among government and NGOs. Specifically, it was mentioned in many interviews that it was important to have a gap analysis between the capacities available and the needs to be fulfilled. Capacity encompasses the response capacities of the responding communities and professionals and the coping capacity of the affected community. Knowing how to protect one's livelihood (such as agriculture, fishery, and hand looming) increases the coping capacity. Interviewees mentioned, for example, the importance of knowing when to harvest just before the flood arrived and which crop to cultivate when the flood started to recede. Similarly, it was important to know how well the local market was still functioning. Key is also a readily accessible

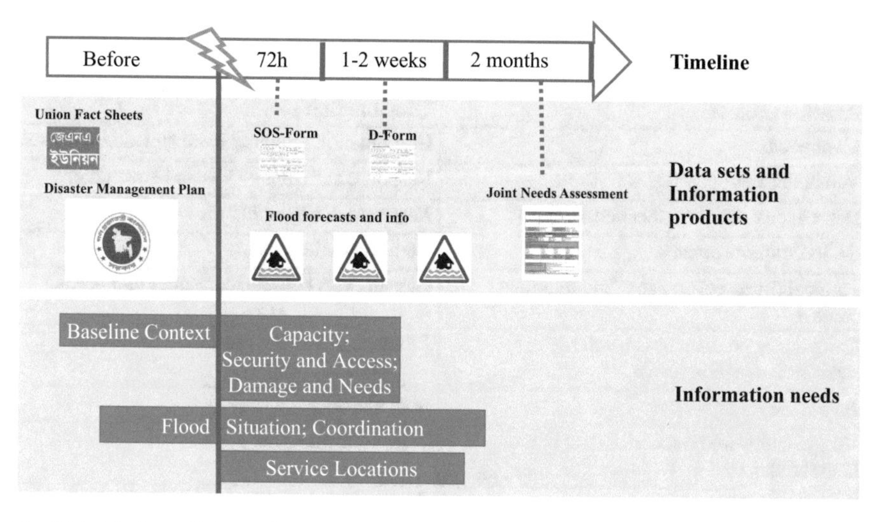

Fig. 18.1 Timing of information needs and availability of data sources

and suitable emergency stock (IFRC 2015). Specifically, information about boat capacity was mentioned as a need in the interviews. The *Baseline* cluster focuses on the context of people before the disaster hit. *Flood news* groups the needs in relation to the arrival and duration of the flood. The *Location Services* cluster refers to locations for essential services such as water, health, food, and shelter, but also to places where there are opportunities for labor. We note however that also for many of the other information needs, a geospatial attribute adds value. *Security and access* refer to access for the responders to the affected community.

(b) **Data sets and information products**

Figure 18.1 depicts the different flood-related data sets and information products, and at which point in time, they were collected and became available. The data sets and information products contain on the order of 40–60 indicators per source at a high level. For example, the Joint Needs Assessment consists of an extensive excel file that compiles answers to 62 questions for several unions (the data set) and an information product (a pdf report discussing and describing the survey results). We did not yet include in our analysis data sets or information products that were available only in Bengali. However, to our current understanding, based on the interviews and our literature and Internet search, this seems to be a minor fraction. We also did not include social media data, given that there is nearly no social media penetration among the affected communities. In the whole of Bangladesh, the percentage of people using the Internet is 9.6% in 2014 (ITU 2015) and most of these are in dense urban areas.

Table 18.2 List of information needs

Crisis impact	Operational environment
Baseline context	**Coordination**
Livelihoods	Coordination groups at local and national level
Vulnerabilities	Response activities NGOs and government
Hazard identification (location, timing)	*Response activities private sector*
Socioeconomic context	Community leaders
Political (local governance) and religious context	Gap analysis between capacities and needs
Community preparedness (such as security/evacuation plans)	Presence of NGO workers
Preparedness of people	Staff skills
Village and ward boundaries (location of households)	Telephone numbers
Damage and needs	*Communication channels*
WASH needs	*Incidents registration*
Health needs	*Evacuation routes*
Education needs (closed schools)	**Capacity**
Food security needs (stoves, firewood)	Stock of emergency items
Shelter needs (including nonfood items)	Coping mechanisms of affected communities
Needs of subgroups (elderly, children)	Local agricultural and fishery situation
Number of people affected	Local market situation
Livestock affected	Institutional capacity
Type of damage to houses	Staff skills and training
Number of damaged houses and number of destroyed houses	Burying strategies
Losses of private belongings	**Service locations (during the flooding)**
Number of people dead and number of people injured	Shelters for humans
People in need of rescue	Shelters for cattle
Submerged houses	Doctors
Damage to infrastructure, health facilities, public buildings	Medicine distribution points/shops
Affected medical personnel	Food buying and selling places
Number of people saved	Labor opportunities
Displaced people	Drinking water locations
Impacted area	Emergency items
Flood situation	*Meeting points*
Flood news	*Pickup points*

(continued)

Table 18.2 (continued)

Crisis impact	Operational environment
Flood duration	**Security and access**
Earlier predictions	News
Time of inundation	Accessibility
Inundated area	*Security*
Drainage and irrigation systems	*Mobile phone coverage*
Flood trend analysis	
Water quality	
River embankment erosion	

The data sets have different levels of collection and aggregation. The government works with the SOS and D form for damage and needs assessments. The union secretary or chairman has to fill in the 10 questions of the SOS form within 48 h after the disaster. The D form has 30 questions, which are filled in and submitted within three weeks. The system is still largely a paper-based system, whereby forms are manually summarized at each of the administrative levels, before they are passed on to central level and digitized. This means that it is often not possible to go back to the data at ward level. This type of data granularity loss we encountered in more data sources. The downward arrow in Fig. 18.2 depicts this risk of data granularity loss at each step up in the government hierarchy. Important data providers are the Department of Disaster Management of the Government of Bangladesh and the Humanitarian Coordination Task Team (HCTT), consisting of UN, NGO, and government representatives. For each file, we singled out all the indicators and determined the data type (excel sheets, relational databases, PDF, text, websites, and geographic information).

(a) ***Mapping data sets and information products on the information needs***
Following the methodology described in Table 18.1, we mapped the 71 information needs on the 15 data sets and information products. We can draw the following conclusions per cluster level, where for now we do not look at time constraints. *Service locations* are not well covered at all with a cumulative coverage of 0.7%.[1] One of the reasons for this might be that information is often collected by phoning people and asking them to give an overview for their ward or by conducting a paper-based survey. We did not come across local responders that use an app or GPS to map locations during the floods. *Capacity* was also not well covered, varying from relatively easy to monitor capacities such as the number of boats up to the more difficult to assess coping mechanisms of affected communities. *Damage and needs* were covered largely by only two out of the 15 data sets (JNA and D form). The following data sources match well the information requirements: JNA (38%), D form (34%), District Disaster

[1]Number of times total coverage of service location information needs by data sources divided by (the number of information needs within service locations) × (the number of data sources).

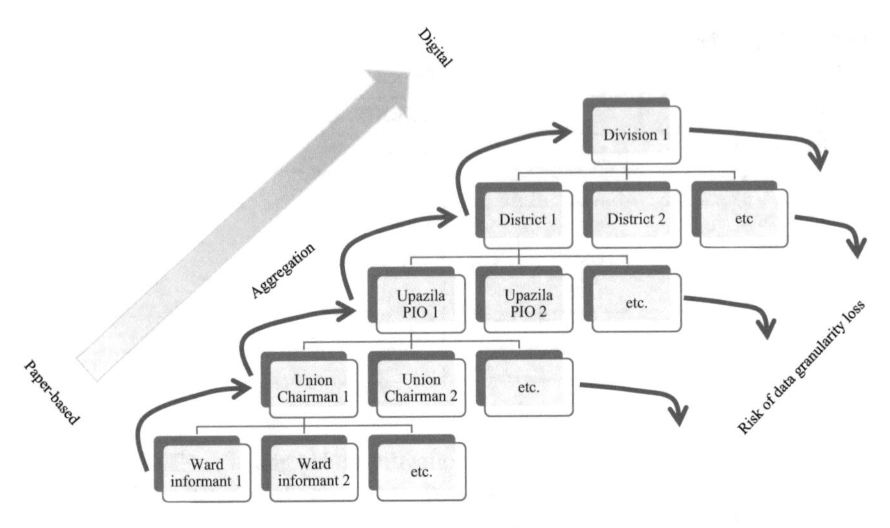

Fig. 18.2 Data collection process within the government for damage and needs assessments with risk of data granularity loss

Management Plan (20%), the (online) News (14%), and the Situation Report (13%). We note that the 13% is based on the first Situation Report that became available; later ones yield a higher coverage. These data sources overlap on some indicators (and hence also their coverage of information needs), and the coverage percentage cannot be summed.

Overall, 75% of the information needs can be covered (49 information needs are fully covered and eight information needs (which were not fully covered) are partly covered by one or the other data set). We also determined which combination of data sources would yield the highest coverage, starting with the highest individual coverage and stepwise selecting the next data source based on its additional coverage. This resulted in a coverage of 68% for the following data sources: JNA, District Disaster Management Plan, and the FFWC website (44 of the 71 information needs are completely fulfilled). None of the data sources fulfills a critical mass of information needs to justify a 100% focus on one source. But an important conclusion is that a very good coverage of information needs can already be reached by the three most important data sets out of the total 15. However, we recall that the above results are without taking timing constraints into account, whereas our interviewees explained that they need certain response information within 48 or 72 h after the disaster strikes. If we do include the timing constraints, then it becomes apparent that most operationally related information is not available in time. In this case, 28.5 information needs are covered, or in other words 40%. Only 27% of all needs are covered in time, whereas 75% are covered if we do not take any timing constraints into account.

18.4 Discussion and Conclusions

We compared our framework of information needs with the one from Gralla et al. (2015). Their *Context and scope*, *Coordination and Institutional Structures* and *Humanitarian Needs* themes overlap with ours and are the most important factors in the earlier response. Several other information requirements are not mentioned in our interviews such as *Looking forward* and *Relevant laws and policies* as part of *Coordination*. The Gralla et al. framework emerged from consultation with mostly responders from the international humanitarian community, whereas our framework emerged from consultation with only national and local responders. Also, the type and scale of disasters looked at was different. We looked at small-scale disasters, whereas Gralla et al. focused on large-scale disasters, where international response is requested by the nation affected. Floods as in our case study have severe impacts on livelihoods but usually less in terms of loss of life. In many cases, there can be also a difference of opinion between the NGOs, on the one hand, and the government, on the other hand, as to whether declare a flood an official disaster. One interviewee mentioned encountering in some cases political pressure not to help. Nevertheless, it is widely acknowledged that the role of national and local responders is of utmost importance also in large-scale disasters. Local responders have more local context knowledge and -in case of recurring disasters like annual floods—they also usually have more response experience than the international community. This leads to a different level of information needs regarding the *Baseline* theme between local, national, and international responders (van den Homberg et al. 2014). For international responders, the *public and media perception* turned out to be a separate theme. In our interviews, media perception did not come forward as an important issue, probably related to the fact that national and local responders usually are not directly applying for funding themselves (but through their supporting international NGOs) and that the local communities affected do not have access to a lot of media channels. We did not find much information needs in relation to *Recovery*. This might have to do with the relatively limited possibilities for the responders in our interview group to extend their activities beyond response. In sum, it is important for each type of context and hazard to develop a tailor-made information needs framework. We have developed one for a hydrological hazard in one of the poorest countries in the world. A comprehensive framework with a generic set of themes can be used as a starting point and for each actor there will be differences as to which category is the most important to them given their organizational mandate, where, for example, some NGOs focus on women empowerment and others on disability. Such a comprehensive framework should include both the local, national, and international perspective.

Subsequent mapping of available data sources on the information needs in the framework is key for identifying the data gaps that currently exist. It is clear from the mapping we did that both the responding and the professional community lack information to effectively dimension and target their response. The governmental SOS form is rapidly available (within 48 h) but is very high level. The more detailed information from the D form is available within three weeks, but the lower government levels often do not have clear guidelines and resources for adequate data collection. NGOs that are part of the Local Consultative Group often do their own assessments, such as in 2014 via a Joint Needs Assessment, creating in fact a new process with different indicators that is only aligned with the government process to a very limited extent. Once the information is collected at central level, support is mobilized for the response, making the response largely a top-down mechanism. Both the NGO and government information architecture are not specifically geared toward coordination and action planning at Community, Union, and Upazilla level, forming a stumbling block for effective local response. To tackle the issues mentioned above, data preparedness activities should become an integral part of the preparedness phase.

First, we propose to organize regular multi-institutional mapping cycles of data sets on information requirements. These cycles should not only consist of keeping an up-to-date inventory of available data sources and providers, but also of regular consultations with responders as to what their information needs are. When the interviewees validated the information needs framework, this sparked their creativity. We got reactions like: "wow, if this is possible, we could also really benefit from X information". It is important hence to keep on evolving the requirements and to use these requirements to shape the information products that providers are creating so that they meet the decision-maker's needs. These mapping cycles will also benefit from advances countries make in terms of open data. Open data can promote inclusion and empowerment, as it has the potential to remove power imbalances that result from asymmetric information, and can give marginalized groups a greater say in policy debates (Davies and Perini 2016).

Second, coordination needs to be improved. A Coordinated Data Scramble (Campbell 2016) can be a very effective way to reach a higher level of coordination in the data collection process, avoiding duplicates, increasing quality, and promoting coherence. It basically means having a multitude of organizations use collaborative platforms and closed digital communication groups for "bounded crowdsourcing" (Meier 2015). Also, specific platforms for managing and sharing the different data sets can be used. Geodash, making use of Geonode, is such a collaborative geospatial platform that was the started up by the World Bank and is now taking over by the Government of Bangladesh (Geodash 2017). UN OCHA deploys the Humanitarian Data Exchange (HDX), more specifically targeting humanitarian data (Keßler and Hendrix 2015).

Third, to facilitate the sharing and exchange of data, standards are being developed and used—to varying degrees—ranging from P-codes for unique geographic identification codes up to the Humanitarian Exchange Language (HXL). Lastly, it will be key to develop capacities of the different stakeholders in parallel to the above activities enhancing their data literacy and access to digital technologies. Especially at the local level, many respondents were, for example, not aware of all the existing data sets nor were they trained in data collection and analysis.

18.5 Future Research

Our research focused on the relation between available data and information needs. Although we inventoried Decisions, Activities, and Information needs, we did not investigate the relationship between these three elements into depth and the system dynamics between the different stakeholders including the political and financial dimension. These dimensions played out, for example, in the still largely separate data collection processes between NGOs and government and in when a flood is declared an official disaster. Further research could address humanitarian decision-making in terms of "what may be influencing decisions, other than the needs on the ground" (Nissen 2015). A political analysis of the stakeholders and the financial flows might strengthen the information management research approach. Regarding the relation between available data and information needs, it will be worthwhile to determine the time dependency of the information needs into more detail and to do the mapping on the data products in a more automated fashion. For large organizations, it is possible to map through which information channels (email, mobile, fax, and chat) information consumers get information products from internal information producers. This kind of mapping does, however, not consider the degree to which information needs are covered. Furthermore, it is much more difficult to do this kind of mapping between organizations and even more so if certain workflows are still paper-based. It might be possible to log data file usage on the main websites that are used by responders and, for example, how the app and dashboard are used (Pachidi et al. 2014). In addition, an after-action review with the responders in a focus group setting could be used to have the responders categorize their needs according to the four phases. This refinement could lead to an enhanced understanding of the data gaps. We envision two avenues to further close these gaps. The first avenue consists of assessing how Artificial Intelligence for Disaster Response (AIDR), such as data and text mining, can be used to link and integrate disparate data sets and to in this way reach a higher coverage of information needs (Spruit and Vlug 2015). It will not be necessary to integrate all disparate data sets; we showed that a very good coverage of information needs can already be reached by integrating the three most important data sets out of the total 14. One could set up so-called data spaces which are loosely integrated sets of data sources where integration happens only when needed (Hristidis et al. 2010). This could become an essential extension to the earlier mentioned data exchange platforms so that these platforms offer—to a certain degree—sensemaking of all the data sets that are shared through them. The second avenue consists of tackling the lack of local and timely data. The Government of Bangladesh has started to develop an online process of collecting the SOS and D form data, tackling in this way the data granularity loss. We have co-created a smartphone application in Bengali that local disaster management professionals and volunteers can use to collect data just before and during the floods that fulfills the currently not covered information needs. The functions and features of the app and dashboard reflect the different clusters of information needs that we identified. The data collected is fed back to the affected communities through a dashboard that is accessible on the very same smartphone. Ultimately, we aim at replicating the same approach also to other flood-affected countries in Asia.

Acknowledgements This research was supported by Cordaid as part of the People Centered Interactive Risk and Information Gateway project and TNO offered internship support to Robert Monné. The authors gratefully acknowledge Marlou Geurts, MA, the program leader from Cordaid for her invaluable support and our interviewees who made so generously time available for providing us with their insights.

References

Campbell, H. (2016). *Coordinated data scramble*. https://www.humanitarianresponse.info/en/topics/imwg/document/coordinated-data-scramble-ppt30june2016. Accessed November 15, 2017.

Comes, T., Walle, B., Brugghemans, B., Chan, J., Meesters, K., & van den Homberg, M. J. C. (2013). *A journey into the information Typhoon Haiyan Disaster: Resilience Lab Field Report findings and research insights: Part I-Into the Fields*. The Disaster Resilience Lab.

Davies, T., & Perini, F. (2016). Researching the emerging impacts of open data: revisiting the ODDC conceptual framework. *The Journal of Community Informatics, 12*(2).

Geodash. (2017). https://geodash.gov.bd/. Accessed November 15, 2017.

Gralla, E., Goentzel, J., & Van de Walle, B. (2015). Understanding the information needs of field-based decision-makers in humanitarian response to sudden onset disasters. In *Proceedings of the 12th International Conference on Information Systems for Crisis Response and Management (ISCRAM)*.

Hallegatte, S., Vogt-Schilb, A., Bangalore, M., & Rozenberg, J. (2016). *Unbreakable: Building the resilience of the poor in the face of natural disasters*. World Bank Publications.

Hristidis, V., Chen, S. C., Li, T., Luis, S., & Deng, Y. (2010). Survey of data management and analysis in disaster situations. *Journal of Systems and Software, 83*(10), 1701–1714.

IFRC Emergency Items Catalogue. http://itemscatalogue.redcross.int/. Accessed November 15, 2017.

Keßler, C., & Hendrix, C. (2015). The humanitarian exchange language: Coordinating disaster response with semantic web technologies. *Semantic Web Journal, 6*(1), 6–21.

Leidig, M., & Teeuw, R. M. (2015). Quantifying and mapping global data poverty. *PLoS ONE, 10*(11).

Meier, P. (2015), Digital humanitarians, how big data is changing the face of humanitarian response. Taylor & Francis Press.

MIRA, Multi-Sector Initial Rapid Assessment Guidance—Revision July 2015, https://www.humanitarianresponse.info/en/programme-cycle/space/document/multi-sector-initial-rapid-assessment-guidance-revision-july-2015, Accessed November 15, 2017.

Monné, R. (2016). *Determining relevant disparate disaster data and selecting an integration method to create actionable information*. M.Sc. thesis, Utrecht University.

Nissen, L.P. (2015), Keynote III Wag the Dog—Information management and decision making in the humanitarian sector. In Palen, Büscher, Comes, & Hughes (Eds.), *Introduction Proceedings of the ISCRAM 2015 Conference—Kristiansand*, May 24–27.

Pachidi, S., Spruit, M., Weerd, I. van der. (2014). Understanding users' behavior with software operation data mining. *Computers in Human Behavior, 30*, Special Issue: ICTs for Human Capital (pp. 583–594).

Preece, G., Shaw, D., & Hayashi, H. (2013). Using the viable system model (VSM) to structure information processing complexity in disaster response. *European Journal of Operational Research, 224*(1), 209–218.

Spruit, M., & Vlug, B. (2015). Effective and efficient classification of topically-enriched domain-specific text snippets. *International Journal of Strategic Decision Sciences, 6*(3), 1–17.

van den Homberg, M., Meesters, K., & van de Walle, B. (2014). Coordination and information management in the Haiyan response: Observations from the field. *Procedia Engineering, 78,* 49–51.

van den Homberg, M., & Neef, M. (2015). Towards novel community-based collaborative disaster management approaches in the new information environment: An NGO perspective. *GRF Davos Planet@Risk, 3*(1), 185–191.

Wahed, A., Rahman, M., Hoque, A., Costello, L., Burley, J., & Walton-Ellery, S. (2014). *Flooding in North-Western Bangladesh HCTT Joint Needs Assessment*. http://reliefweb.int/sites/reliefweb.int/files/resources/0809_NW_Flooding_JNA_FinalFINAL.pdf. Accessed November 15, 2017.

Walton-Ellery, S., & Rashid, H. (2012). Joint needs assessment works: Taking it forward together. http://www.lcgbangladesh.org/HCTT/LLreport_final_121112.pdf. Accessed November 7, 2015.

Wolbers, J., & Boersma, K. (2013). The common operational picture as collective sensemaking. *Journal of Contingencies and Crisis Management, 21*(4).

Printed in the United States
By Bookmasters